Complexity in urban crisis management

Amsterdam's response to the Bijlmer air disaster

COMPLEXITY IN URBAN CRISIS MANAGEMENT

Amsterdam's response to the Bijlmer air disaster

Uriel Rosenthal
Paul 't Hart
Menno J. van Duin
R. Arjen Boin
Marceline B.R. Kroon
Marc H.P. Otten
Werner I.E. Overdijk

Crisis Research Center
Department of Public Administration
University of Leiden
Erasmus Universiteit Rotterdam

earthscan
from Routledge

First published 1994 by James & James (Science Publishers) Ltd

This edition published 2013 by Earthscan

For a full list of publications please contact:
Earthscan
2 Park Square, Milton Park, Abingdon, Oxon OX14 4RN
Simultaneously published in the USA and Canada by Earthscan
711 Third Avenue, New York, NY 10017

First issued in hardback 2017

Earthscan is an imprint of the Taylor & Francis Group, an informa business

A catalogue record for this book is available from the British Library

ISBN 13: 978-1-138-41078-7 (hbk)
ISBN 13: 978-1-873936-34-4 (pbk)

Typeset by Ann van der Zwalmen, Leiden, The Netherlands

Front cover photo: Ad van Denderen

Back cover photo: ANP

Partly translated from the dutch by Anthony Fudge & Associates, Amsterdam, The Netherlands

Contents

Appendices

Introduction

The evening of Sunday 4 October 1992. A nightmare scenario: a Boeing 747 freighter, out of control, bored its way into two apartment blocks, taking the people living there totally unawares. A disaster out of the blue, with nothing to salvage or save for the first few hours. It took a week before the city of Amsterdam and the Netherlands as a whole could sum up the tragedy in a single phrase: 'the Bijlmer disaster', after the name of the well-known high-rise Amsterdam suburb where the crash occurred. It cost 43 lives, including the aircraft's crew of four.

The Amsterdam authorities and city services are used to dealing with sudden problems. But the tragedy that occurred at 18.36 that Sunday evening was to be the start of an extraordinary period of intensive activity and effort that demanded, across a broad front, the utmost the city's people and organisations could possibly give.

In the Dutch context, but possibly also from a wider Western perspective, the Bijlmer air disaster represented a paradigmatic example of the vulnerability of major urban centres, and the complex social, political and organisational issues that this situation raises:

An air crash. Accidents involving aircraft always appeal to the public imagination and by definition attract widespread attention. This relates to more general feelings of danger inevitably prompted by air travel.[1] Despite the relatively low danger of an aircraft accident, and the fact that air travel is safer than, for instance, going by road, the consequences of any such accident are usually instantly fatal. There is no escape from a crashing aircraft; and images of aircraft wreckage stay imprinted on the mind for a very long time.

An air crash in a residential area. The fact that the aircraft, a Boeing 747 – one of the largest types in civil aviation – hit a major housing complex, causing an inferno, was shocking in itself. But when an air crash kills not only the crew and passengers, but also a large number of civilians on the ground, one approaches the 'worst-case scenario'.

To an extent, the fact that air crashes involve the deaths of a large number of passengers and crew has become an accepted risk.[2] Not for nothing are emergency instructions given before every flight. And not for nothing are there strict regulations to the effect that airports must have fire brigades on permanent stand-by. But when unsuspecting civilians die because an aircraft suddenly crashes on their block of flats, we have an 'unacceptable risk situation'.[3] The safety of people living in the neighbourhood is rarely a factor in discussions on siting an airport or building an additional runway. Insofar as the location is a point at issue, it is almost invariably from the angle of noise and other pollution.[4] Here, the only difference from a worst-case scenario was that fortunately the 747 was a freighter, with only one passenger and a crew of three on board.

An apocalyptic scene. Several of our interviewees experienced the Bijlmer disaster as a 'nightmare scenario'. The fact that it happened after dark made the flames even more savage and threatening. With justification newspaper headlines read, '...like something out of a film'.[5] Some of those involved mentioned the bizarre and unreal nature of the disaster.

The Bijlmer of all places. The Dutch are familiar with the Bijlmermeer – the Bijlmer for short – as one of those areas where contemporary social problems have a way of accumulating. For many people in the Netherlands, the blocks of flats so characteristic of the Bijlmer are an object lesson in 'how not to do it' as regards town planning.[6] For many people the very name Bijlmermeer had become synonymous with aggression, crime, drug abuse, urban decay, empty dwellings and other social misery. Slowly, and with great difficulty, the local authorities had succeeded in rechristening the area with the more neutral title Amsterdam Southeast. But, given its reputation, it was no surprise that many people's reaction to the news of the air crash was 'not the Bijlmer *again*'.

Multi-cultural mix. A great many different ethnic groupings were involved in the disaster. From the very start, community organisations represented the specific problems and interests of their people. The multi-cultural nature of the disaster was most visible during the public mourning ceremonies on Sunday 11 October. Each grouping bade its departed farewell in their own way. The colourful and diverse character of the memorial service in the RAI exhibition centre was impressive from every point of view.

Ongoing uncertainty. The Bijlmer disaster continued to hold many in its grip for a long time, due to the extended period of uncertainty as to the exact death toll. Initially it was feared that at least 250 people had died. The large number of missing persons and the many rumours – for example, about a crowd allegedly attending a disco party in the basement being buried under the debris – seemed to bear out this high figure. Later it transpired that far fewer had died. But the high initial figure remained in many people's minds for a long time.

Accumulation of problems. The days and weeks after the crash saw a constant stream of new events and problems arising. Obviously there was considerable interest in the actual cause of the crash: the final minutes of the flight, communications with the control tower, the dragging of ponds and lakes for aircraft debris, the missing voice recorder, the finding of the 'black box'. But, in the meantime, the visible social consequences of the disaster also produced their fair share of dramatic moments: the operation to recover bodies, the memorial service, the problems in providing relief, the hundreds of illegal immigrants outside the Population Register building on one of Amsterdam's main canals. And so the disaster was notable not only for its emergency phase, but also – after several days had passed – for a whole range of equally striking post-disaster developments.

The media. The Bijlmer disaster struck hard and deep, and media coverage was commensurate – which posed some delicate problems. During the weeks following the crash, its tragic consequences remained an important issue, precisely because of the heavy coverage. Day after day, the media devoted hours of broadcasting time and pages of print to stories on the disaster. Some of the more dramatic pictures were shown dozens of times.

Viewed in terms of other emergencies, the Bijlmer disaster shows both typical and atypical characteristics. Typical disaster elements were the surge in solidarity in the first few hours and days, the large numbers of people and agencies descending on both the disaster site and the crisis centres, the extensive informal and spontaneous help offered, the deep complexity of the situation and the speed and variety of rumour-mongering.

But there were also factors that, in a certain sense, made this an atypical disaster. In terms of emergency relief the Bijlmer crash could be broadly classified as a large-scale incident – as evidenced, among other things, by the strikingly few injured. The general rule for a disaster is that there are 5–10 times as many wounded as dead.[7] Here, most of the dozens of ambulances sent to the scene were recalled to base without having moved a single patient. Not one single body was recovered on the first evening, and for the first few days there was genuine uncertainty about the numbers killed.

In a certain sense, the emergency operations to deal with the disaster were of only limited significance. For many organisations the hard work started only on Monday 5 October. Organisations and services were saddled with unfamiliar tasks. On Sunday evening the fire brigade fought the blaze using heavy equipment; in the days that followed they had to work to painstaking levels of precision in the operation to recover human remains. On Sunday evening the police followed standard procedures for large-scale operations; later on they needed a quite different approach. Immediately after the crash, the city Health and Medical Service mobilised personnel and equipment suitable for extensive medical activities; then, in the course of the week, it was charged with the preparation of an after-care plan.

The atypicality of the Bijlmer disaster is also apparent from the most important tasks fulfilled by the relevant authorities and services during the period the disaster declaration was in force. For example, none of the crucial strategic decisions that would determine the fate of the operations during the emergency phase of the disaster were actually taken on the first night. As time moved on, the nature of the city's activities shifted from counter-disaster measures to crisis management.[8] The term crisis management itself indicates the broader administrative and policy-making challenges presented by the extraordinary circumstances confronting the Amsterdam city administration throughout the disaster period. The crash of the Boeing 747 prompted a series of developments in the local community that far outstripped the standard tasks of dealing with a major disaster-type incident, e.g. fire-fighting, securing the area and victim relief. The specific features of the population affected were a major contributory factor: dense, multicultural, socially disadvantaged, complex residential configurations, and relatively high concentrations of illegal immigrants. That complicated the implementation of relief and aid activities and confronted some city officials and organisations with phenomena and problems they had scarcely dealt with before the disaster. Constant press attention and the eventual hardening of the climate of public opinion in the aftermath of the disaster also contributed to the generalised crisis atmosphere.

In many ways, therefore, the Bijlmer air crash was not one but a cluster of disasters. Its location created complex issues. It made the Bijlmer's own special set of problems boil over, forcing the authorities to implement crisis management alongside combatting the disaster. In the days and weeks after the crash the social situation in the Bijlmer threw up intractable problems and dilemmas during the effort to assist people from a rich variety of cultural backgrounds, including groups of illegal immigrants.

Field study

The Crisis Research Center (CRC) was present at the disaster site on the Sunday evening of the crash. Several members of CRC helped however they could in the Bijlmer Sports Centre during the first few hours. In the ensuing days, CRC undertook extensive observations in and around the Bijlmer area.

In December 1992 the Mayor and Aldermen of Amsterdam commissioned CRC to carry out an evaluation study of the Bijlmer air disaster. CRC's assignment was to carry out a broad-based study, which should include a description and analysis of the performance of the Crisis Co-ordination Centre and operational services in the weeks following the crash. The City Council requested that it should be 'a thorough study, identifying lessons and where possible and appropriate improving disaster counter-measures'.

In line with this brief, CRC concentrated throughout on lessons to be drawn from the Bijlmer disaster. This implied a detailed reconstruction and analysis of the events and an evaluation aimed at exposing and learning from the strengths and weaknesses of the city administration's counter-disaster and crisis management activities. The results of international research into counter-disaster measures and crisis management were used in further defining the questions posed in the study and in the collection and reporting of data. The James & James English edition of the study has been edited down from the original report as far as local information was concerned. Also, more detailed evaluations of the performance of each of the services concerned have been omitted. The concluding chapter was written especially for this volume, and attempts to go beyond the case study to look at some more general issues of urban crisis management raised by the Amsterdam experience.

This study has relied on a variety of sources, including Dutch and foreign news publications; all relevant city documents and records, particularly log books and evaluations of city services, the Co-ordination Centre and the Southeast Neighbourhood Council; Dutch television recordings, including news broadcasts and current affairs programmes; extended interviews with some 50 people in and outside the city administration; and a mid-March 1993 telephone survey among the inhabitants of Amsterdam.

Past disaster and crisis studies have taught us that perceptions of events will differ among the individuals involved, and that it is very difficult, if not impossible, to reach a definitive reconstruction of 'what really happened'. We had the same experience during this study. When describing and evaluating a series of complex and shocking events, as in the Bijlmer disaster, it is often simply not feasible – nor, from the angle of scientific accuracy, desirable – always to seek a single depiction of the 'facts', or an exclusive view of the significance of given events. Documents and interviews will sometimes produce diverging reports and interpretations. We by no means rule out partiality and protection of interests here; but, more often than not, different recollections seemed to be more a matter of the location or context from which people were involved in events.

In general, our study shows that the authorities and city services reacted to the disaster and its immediate consequences effectively and with due consideration. Emergency counter-disaster action and care activities went well. The Amsterdam city administration showed itself to be a 'caring government'. But we echo the statement of one of our interviewees that the least spectacular aspect of the counter-disaster activities and crisis management was probably the most difficult, namely the co-ordination of care for the victims and the subsequent furore surrounding the treatment of illegal immigrants. Although the notion from the international disaster literature of the 'disaster after the disaster' may not fully apply in this case, here too social solidarity and compassion began to wane just one week after disaster had struck; and the city authorities and services were not immune to this.

Overview of the study

The broad approach of this study implies that the reconstruction, analysis and evaluation go beyond the immediate counter-disaster and care provisions in the first few days following the disaster, to include the various problems that came in its wake. Yet the scope of the study is confined to the duration of the official disaster period, which formally ended on 27 October 1992.

The disaster (Chapter 1). This describes what happened. The emphasis is on the first few days and weeks. There is a short treatment of the aftermath and of the broader impact of the disaster on the local community. The next five thematic chapters reconstruct and analyse what we regard as the five most essential challenges confronting the city authorities during the period of the disaster; these are presented in summary form below. As time progressed, the last two challenges became more important and meeting them more problematic.

Emergency counter-disaster measures (Chapter 2). The most important emergency services went into action immediately after the aircraft had crashed into the flats. At this acute stage of combatting the disaster, a central role was played by the local fire brigade, supported by neighbouring brigades, and the police. In addition to police, fire-fighters and the medical services, there was also a large and active presence of mainly voluntary relief organisations like the Red Cross and Salvation Army. And, during the course of the evening, dozens of other volunteers reported to the various relief centres. This initial period was notable for fast, large-scale operational activities. In a way, the fact that there were so few casualties and no dead had been recovered made the situation seem unreal. Few strategic decisions were taken at this stage; the operational services 'just got on with the job'. Most significant were the Mayor's declaration of a disaster situation and the decision to centralise incoming and outgoing information.

Recovery, identification and the missing (Chapter 3). From the morning of Monday 5 October to the afternoon of Friday 9 October, considerable administrative and operational effort went into the recovery and identification of bodies, and the search for missing persons. Parts of the buildings had collapsed, but precisely what effects this had had, no one knew. Many people feared that family or friends had died as a result of the crash. The recovery operation and the subsequent identification of remains were meant to clear up the uncertainty; in addition, the missing persons enquiry was also designed to show exactly who had died in this tragedy. The significance of the Co-ordination Centre started to increase around this period. Although the acute danger had passed, the authorities and families and friends still perceived this uncertainty on the death toll with alarm. In turn, this prompted decisions by the authorities to clear up the uncertainty as soon as possible.

Psycho-social relief (Chapter 4). Activities at this stage focused on the survivors and their relations, continuing from the evening of Sunday 4 October for several weeks, and involving many different city and other organisations. Meanwhile, the police continued to play an important role. Several Aldermen, including Mrs Wildekamp and Mr Genet, Mr Jonker and Mr De Grave, were involved in activities such as the memorial service and the planning of funeral arrangements and aftercare. The Bijlmer Sports Centre was at the heart of relief provisions for victims and their families and friends. Relief workers from a range of organisations endeavoured to give support to the victims.

Material relief (Chapter 5). The city complemented the psycho-social relief with various types of material support. This included replacement housing, emergency social security payments, household requisites, and other tangible goods and services. However, as time moved on, most of the organisations involved began to entertain reservations about being too lenient in interpreting the concept of 'caring government'. Within a few days they were confronted by fraudsters – the so-called 'free riders' – trying to obtain material benefits to which they were not entitled. Many were well-known trouble-makers. After the first week, various organisations were obliged to spend more time on separating the real victims from the fraudulent pseudo-victims. Pass systems, several re-registration processes and a range of checks were all designed to achieve this. However, the problems involved in sorting matters out dragged on for months after the crash.

Crisis decision making (Chapter 6). The Co-ordination Centre established at City Hall and presided over by the Mayor fulfilled four key functions during the disaster period, each of which is analysed in detail. It defined the situation in administrative and policy terms, setting the stage for operations and determining the main thrust and duration of the emergency response. Secondly, the Centre provided the key arena for the exercise of political and administrative leadership in co-ordinating the response of the city's apparatus and that of other parties. Thirdly, incoming and outgoing flows of information were controlled here, with a pivotal role in the management of media contacts. Finally, one of the most sensitive responsibilities was to lead the community in coping with the collective stress that followed the disaster.

Conclusion (Chapter 7). In the final chapter we look back at the Bijlmer air disaster within the wider context of emergency management in urban environments.

Appendices. The appendices at the back of this report provide a chronology of events, brief results of the public opinion poll, a summary of published sources and a list of groups and organisations whose representatives we interviewed during the course of this study.

Acknowledgements

We greatly appreciate our interviewees' frankness and willingness to co-operate. Everyone we approached was prepared to make time available for an interview. For some of them, recalling the terrible events was difficult – on an emotional as well as a practical level. In addition, relevant documents were made available to us by the city services, the Southeast Neighbourhood Council and several other organisations within and outside Amsterdam. Our thanks also go to the Information Directorate of the Dutch Ministry of Transport and Public Works for providing us with comprehensive media overviews, and to the Dutch Ministry of the Interior for financial assistance in preparing this international edition. Finally, CRC is most grateful to Prof. Dr. H. Daudt and Dr. G.H. Scholten for their critical commentaries on earlier versions of this evaluation.

Notes

1 See, for example, C. Perrow (1984). *Normal Accidents: Living with High Risk Technologies*, Basic Books, New York.

2 Sources for a detailed treatment of the concept of 'acceptable risk' include B. Fischoff, S. Lichtenstein, P. Slovic, S. Derby and R. Keeney (1981). *Acceptable Risk*, Cambridge University Press, New York.

3 Perrow (see note 1) distinguishes between categories of victims. 'First-party' victims comprise workers present at the moment of the disaster, or the aircraft's crew. 'Second-party' victims are also directly involved in the disaster, but cannot influence it (e.g. passengers). 'Third-party' victims have no involvement whatsoever with the system. (The Bijlmer flat-dwellers belonged to this category.) Lastly, Perrow distinguishes 'fourth-party' victims: the as yet unborn generations of victims of disasters involving harmful/toxic substances.

4 E.J. Feldman and J. Milch (1982). *Technology Versus Democracy: The Comparative Politics of International Airports*, Auburn House, Boston.

5 *De Telegraaf*, 5 October 1992.

6 See M. Mentzel (1989). *De Bijlmermeer als Grensverleggend Ideaal (The Bijlmermeer as a Path-breaking Ideal)*, Universitaire Pers, Delft.

7 *Blauwdruk voor de Unité de Doctrine in de Rampenbestrijding (Blueprint for a Uniform Philosophy of Disaster Management)* (November 1986), Ministry of The Interior, The Hague.

8 See P. 't Hart and U. Rosenthal (Eds) (1990). *Kritieke Momenten: Studies over Beslissen in Moeilijke Omstandigheden (Critical Moments: Studies on Crisis Decision Making)*, Gouda Quint, Arnhem.

CHAPTER 1

The Bijlmer disaster – the events

Sunday evening, 4 October 1992: the crash and the initial response

Flight LY *1862*

Just before 18.30 on the evening of Sunday 4 October 1992 an El Al Boeing 747 took off from Schiphol, the main airport in the Netherlands (located near Amsterdam). The four people on board the Israeli freight plane, three crew members and a passenger, were on their way to Tel Aviv.

The first few minutes of the flight were uneventful, but shortly after take-off the crew were confronted with technical problems. In his first alarm message to the Schiphol control tower, the captain reported that both starboard engines had stopped functioning. At this time the crew were probably unaware that starboard engines 3 and 4 had become detached from the wing and had fallen into a lake.[1] Seemingly calm, Captain Fuchs requested permission to return to Schiphol airport. Only minutes after the plane's departure, the captain suddenly announced that he was experiencing problems in controlling the aircraft. The last message from the flight deck followed seconds later: 'LY 1862, going down, going down'.[2] At 18.36 the Boeing 747 freighter disappeared from the radar screen.

The 747 crashed into a neighbourhood known as 'the Bijlmer', situated in Southeast Amsterdam. The aircraft plunged directly into two, linked blocks of 10-storey flats, named 'Groeneveen' and 'Kruitberg', precisely at the corner where one joined the other. The damage caused by the plane's impact was enormous. Thirty-one flats were demolished at a single blow, causing a gaping hole where before the apartment buildings had connected. Almost instantaneously, flames engulfed the apartments neighbouring the ones hit by the plane. Fed by the fuel from the aircraft's tanks, the inferno consumed another 49 flats. The apocalyptic impression of the disaster site was reinforced by the ground fires blazing everywhere in the wreckage of the aircraft and the buildings. Whipped by a strong wind, thick columns of smoke rose up, masking the crash area. The combination of impact, smoke and fire damage made a total of 266 apartments uninhabitable.[3] Later, part of the complex would have to be demolished.

Immediately after the crash, there was total chaos in and around the two buildings now separated by a burning pile of rubble. People living in adjacent units managed to escape from the blazing buildings, while others entered the buildings in an attempt to reach and rescue possible casualties. Of the final death toll of 43, most had lived in the 31 flats hit directly by the aircraft. The dead also included the El Al crew and their passenger.

Raising the alarm

The operational services. For the operational services involved, it was clear from the start that this was a major incident. Despite initially contradictory reports on the location of the crash, within a few minutes the police established from incoming calls that the aircraft had come down on these two blocks of flats in the Bijlmer. Personnel and equipment from various disaster teams were mobilised immediately and directed to the disaster site: the first ambulance was on its way at 18.37. Taxi drivers volunteered to take casualties to nearby hospitals, at no charge. The central reception point for medical assistance was the Academic Medical Centre (AMC) – the university hospital – where the disaster plan was put into operation at 19.14. Hospitals in nearby communities were on stand-by.

Firemen at the local fire station had witnessed the aircraft crashing into the apartment blocks and arrived very shortly after with the two available fire trucks. Other fire departments, including Schiphol Airport's specially equipped crash tenders, also made for the site of the disaster. Within 90 minutes of the crash almost 300 fire-fighters were in action at the wrecked buildings.

Ten minutes after impact, dozens of police officers were on their way to the scene. Both on the spot and in the vicinity, the police soon had their hands full keeping out sightseers and sealing off the disaster area. During the course of Sunday evening, some 500 officers were deployed, including motorcycle, mounted and riot squad units.

The authorities. The Mayor of Amsterdam, Ed van Thijn, was at home watching his favourite soccer team Ajax play on TV when he received a call from the head of the city's Information Department. She told him she had heard that an aircraft had crashed in the Bijlmer. The Mayor's first reaction was disbelief.[4] When the Chief of Police, Nordholt, confirmed the report some minutes later, Van Thijn realised that a catastrophe had just occurred in his city. At around 19.15, the Mayor issued a verbal statement declaring the situation a disaster.

Following the guidelines of the Amsterdam disaster plan, the Co-ordination Centre went operational. Located in a bomb-proof shelter under City Hall, the Co-ordination Centre was specially constructed for times of major emergency; it is equipped with modern communications equipment and can accommodate a large group of people for lengthy periods. It was here that the Mayor and his officials assembled. Most of those who had been alerted arrived at City Hall before 20.00.[5]

During the first hour efforts were devoted to establishing direct lines of communication with the disaster site. While the Co-ordination Centre filled up with people, the Head of Public Order and Safety acted as Co-ordinator for the Centre. In this position she introduced newcomers to the Co-ordination Centre and allocated their tasks. Once they had found their places, these officials tried to make contact with their own services. During the first few hours after the crash, activity in the Co-ordination Centre focused on getting a picture of the scale of the disaster, while at the same time work got underway on organising counter-disaster measures (see the map on p. 12 for an overview of locations of crisis centres).

Emergency services

At the disaster site, fire-fighters were busy dousing proliferating ground fires and containing the spread of flames in the stricken buildings. The raging inferno hampered the attempts of rescue workers to reach 'Groeneveen' and 'Kruitberg'. Despite this, fire-fighters managed to search several sections of the two buildings. No survivors were found. Meanwhile there was a constant risk of the buildings collapsing and the added danger of explosion from gas mains and aircraft parts, so the gas mains were sealed off during the course of the evening. Following the crash, a growing stream of people moved towards the disaster area, creating heavy pressure on the temporary police lines around the site. Initially, these 'disaster tourists' could simply walk through at certain points but, two hours after the disaster occurred, the police started to clear people away from the area. Major traffic jams clogged roads in the immediate vicinity as routes leading to the disaster site were largely blocked by the inflow of spectators and people trying to find out whether their friends or relatives had been hurt.

Despite the traffic chaos, fire-fighting progressed well. Just over an hour after impact, the wreckage of the aircraft and the buildings were covered by a blanket of foam, and the danger of explosion dissipated. At 21.07, two and a half hours after the aircraft had crashed, the fire service announced that the fire was under control, although it would be 22.30 before the most important seats of the fire had been extinguished. Under the floodlight of hovering helicopters, a fire brigade rescue squad prepared to go into the ruins.

First aid on the spot

Such was the enormous damage caused by the impact of the El Al 747 that a high death toll appeared likely and persistent rumours circulated to this effect. The ANP news agency reported that the city Medical Services had found the bodies of

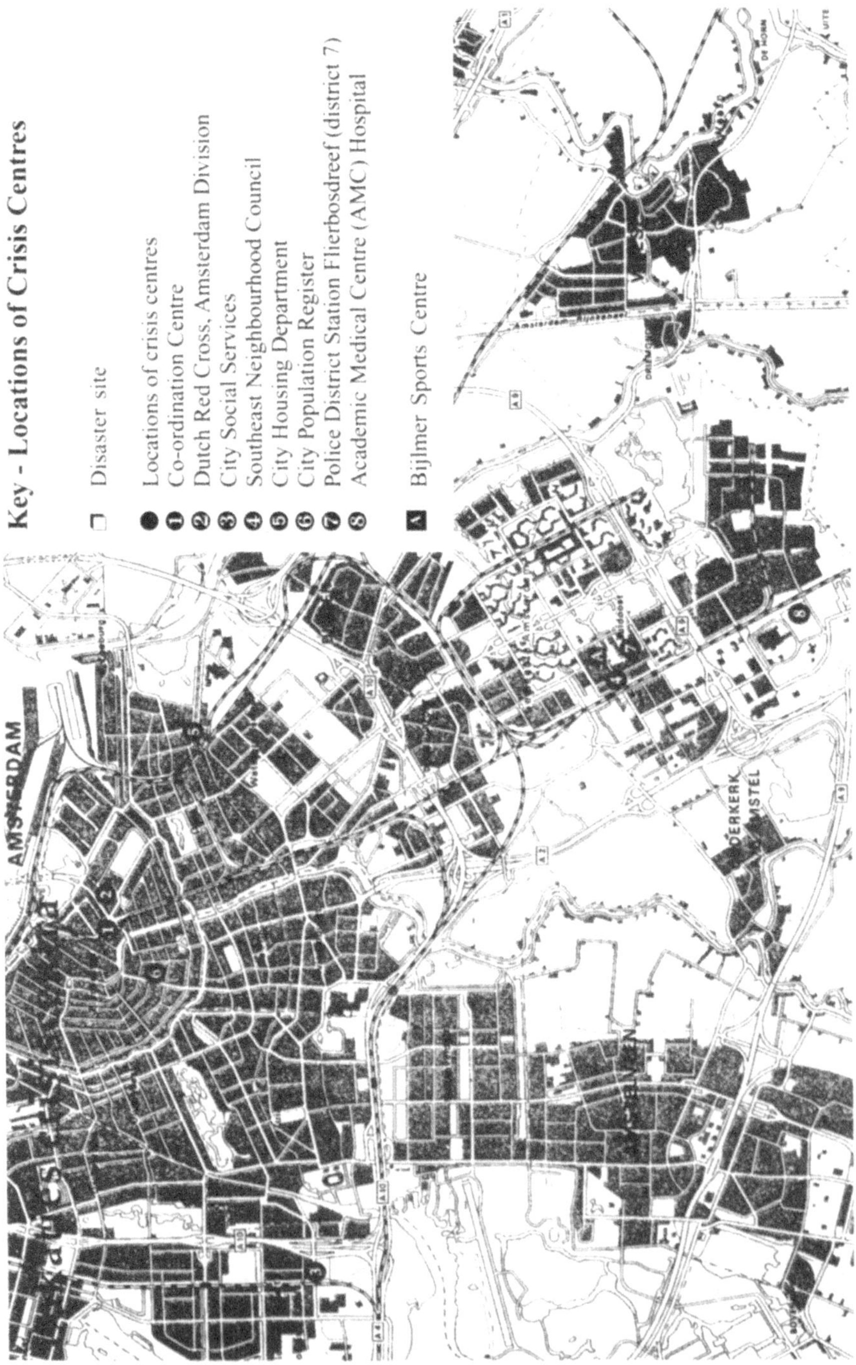
Key - Locations of Crisis Centres
Disaster site
Locations of crisis centres
1 Co-ordination Centre
2 Dutch Red Cross, Amsterdam Division
3 City Social Services
4 Southeast Neighbourhood Council
5 City Housing Department
6 City Population Register
7 Police District Station Flierbosdreef (district 7)
8 Academic Medical Centre (AMC) Hospital
A Bijlmer Sports Centre
AMSTERDAM

dozens of people who had jumped to their deaths from the higher floors. According to other rumours there were a large number of people in the buildings at the moment of impact. For several hours the media and members of the Co-ordination Centre were under the impression that 12 bodies had been recovered.

Medical teams, however, had relatively little work to do for the first few hours and at around 22.00 the Central Ambulance Station withdrew several dozen ambulances. By 22.40 a total of 36 wounded had been taken from the disaster area to nearby hospitals.

A large number of people needed help and support. Some of them were looked after at the disaster site, and were then taken by buses to one of the emergency relief locations. Other people were able to reach these locations on their own. On the Sunday evening, several places in the Bijlmer were spontaneously converted into relief stations for victims and local people but during the course of the evening, the spread of victims over several reception centres began to fragment relief activities. Hence, at around 23.30, the Co-ordination Centre – at the suggestion of the Southeast police district – decided that the Bijlmer Sports Centre would be the central reception point for all victims of the disaster.

The people in other reception centres were then bussed to the Sports Centre, where voluntary workers, city services and professional organisations assembled to provide relief and assistance. The police guarded the entrance to the Sports Centre and registered everyone who entered. The hall was besieged by camera crews and journalists, who were duly removed from the complex by the police. At around 03.00 on Monday morning, a large number of now homeless people were transferred from the Bijlmer Sports Centre to overnight accommodation elsewhere. Most of the approximately 200 homeless were taken to nearby naval barracks. Other overnight accommodation was arranged in the Social Services Department's emergency facility. The relief workers stayed behind in the Bijlmer Sports Centre, together with some of the victims.

The Co-ordination Centre: initial activities

At first the Co-ordination Centre lacked any overview of developments in the disaster area. The scarce information available came from Fire Chief Ernst and Chief of Police Nordholt. It was known that the aircraft was not a passenger-carrying type, but there was minimal information on the dead and wounded. There was also a lack of clarity on the number of apartments affected. Once information had been obtained on the number of units involved, the City Population Register was directed to provide details on the number of people registered in the relevant units.

The Bijlmer air disaster was a shocking event, prompting reactions from around the world. City Hall was swamped by press representatives. Press conferences were held in City Hall, at 21.30 and 23.00. Despite the lack of available information, the national and international press was kept up to date as far as possible. At 22.30 Mayor Van Thijn, Fire Chief Ernst and Chief of Police Nordholt went to the site of the disaster to size up the situation for themselves. At midnight they gave together what was to be the first of a long series of press conferences.

From their position under City Hall, the Amsterdam authorities had regular contact with their counterparts at the provincial and national levels. After the crash, the Ministry of the Interior in The Hague had manned its National Co-ordination Centre, to assist the Amsterdam authorities if necessary. The North Holland provincial authorities established a crisis centre at Haarlem, where the Provincial Governor of North Holland, Van Kemenade, was kept informed on the situation in Amsterdam. Ministers Ms Dales (Interior) and Mrs Maij-Weggen (Transport and Public Works) were also briefed.

The first week

Recovery operations commence

Operations to recover human remains commenced at dawn on Monday morning. Precautions taken during the night ensured that conditions were safe enough for the rescue squads to enter the ravaged buildings. A mobile crane was used to remove loosely hanging concrete panels, one by one. Several different services were involved in the search through, and removal of, the rubble: the Amsterdam fire brigade and police, regional fire brigade units, members of the National Police Disaster Identification Team (the RIT), aviation experts, civil defence volunteers and various private contractors. The investigation into the cause of the crash ran parallel with recovery activities and concentrated on the missing flight recorder (the 'black box').

But, primarily, the operation was designed to recover and identify human remains. Recovery got off to a slow start as RIT's painstaking approach took time. and the rubble of the flats and the aircraft wreckage could only be removed piecemeal. The pace of the recovery operation was largely dictated by the time-consuming process of identification. Other problems hindered the recovery operations. Small fires kept springing up in the ruins of the two buildings; smoke also hampered the teams' work, while the risk of collapsing masonry remained.

Organisation of relief and assistance

On Monday, buses shuttled between the various places offering overnight accommodation and the day-time facility at the Bijlmer Sports Centre. In the course of the morning, many survivors of the crash collected in the sports hall. All those affected were registered on arrival. People searching for family and friends were also admitted to see if perhaps someone they knew might be there.

Several aid organisations offered their services in the Bijlmer Sports Centre. Input from bodies like the city Department of Social Services, the city Housing and Health Service departments, the Red Cross, the Salvation Army and the Insurance Association started on Monday 5 October. Health Department staff were busily providing a wide range of assistance to victims. The Health Service was also active in organising and equipping the sports hall and the overnight facilities: this included co-ordinating the transport of those involved between the Bijlmer Sports Centre and the overnight accommodation.

Besides the Southeast Neighbourhood Council, various immigrant organisations were represented in the sports hall. By Monday it had become clear that a relatively large number of immigrants had been made homeless by the disaster. The Surinamese, Ghanaian and Antillean communities in the Bijlmer appeared to have been particularly badly hit.

On Monday morning the various municipal services started to organise longer-term relief for those affected. The Housing Department contacted the Nieuw Amsterdam Housing Co-operative, which owned most of the flats and houses in the Bijlmer, to seek new accommodation for the survivors. Other housing co-operatives were also asked to look for additional space. The housing co-operatives reacted swiftly: routine allocation of homes was halted and all accommodation falling vacant was reserved for people made homeless by the disaster. By Monday it was already clear that there would be sufficient accommodation available for the survivors. However, most of these homes were not immediately available; often they needed to be decorated and furnished before people could move in.

On Monday morning, Housing Department staff began drafting the criteria whereby people would be allocated new homes, and drawing up a set of policy guidelines as to who would be eligible for new housing. Based on these rules, the Housing Department laid down who would and would not be considered victims of the disaster. At the same time, the Housing Department and the Nieuw Amsterdam Housing Co-operative desks at the sports hall started to compile lists of applicants for accommodation.

Royal visit to the disaster site

On Monday morning, Queen Beatrix visited the area of the disaster. It was the first official visit to a disaster site of the Queen's reign (which started in 1980). The fact that on this occasion she had immediately decided to go to the Bijlmer was indicative of the shock and horror felt throughout the Netherlands. The Queen was accompanied by her son, Crown Prince Willem-Alexander, a delegation of cabinet ministers headed by Premier Lubbers, the Provincial Governor of North Holland, Van Kemenade, and a delegation from Amsterdam City Council, led by the Mayor. It had been previously decided at the Co-ordination Centre that all the dignitaries should come at the same time and be shown round as a group, the idea being to hinder the recovery teams as little as possible. After a short visit to survivors and their families in the Bijlmer Sports Centre, the group was taken to the wrecked apartment buildings. The Queen was visibly moved at the sight of the rubble and the gaping hole; other dignitaries were also clearly emotionally affected. At a press conference after the visit, Premier Lubbers called it 'an awful drama'.[6]

Managing the recovery operation

The co-ordination of the counter-disaster operations and relief to the victims was led by Mayor Van Thijn. In the Co-ordination Centre, information was constantly collated and interpreted as it came in, while the provincial and national authorities, as well as the Dutch and international media, were periodically briefed. As Mayor Van Thijn put it in a special council meeting on Monday afternoon, the pace of the recovery operation was 'exasperatingly slow'.[7] Twenty-four hours after the plane had come down, only six bodies had been recovered. By Monday evening, more than 600 persons had been reported missing. The discrepancy between the small number of bodies recovered and the much larger number of missing persons seemed to suggest that there would be many more dead beneath the rubble. Moreover, in the Co-ordination Centre it was suspected that not all casualties could be traced in the Population Register, particularly since there might well have been many illegal immigrants living in the flats at the time of the crash.[8] In addition, there were still persistent and hard-to-pin-down rumours that people who just happened to be in the area that Sunday night had also fallen victim to the crash. According to Mayor Van Thijn, speaking in the special City Council meeting, more than 250 people were missing and 'alas, there is not much hope that they are still alive'.[9]

In the days immediately following the crash, members of the Co-ordination Centre team attempted to get an overview of the total death toll as fast as possible. On Monday evening it was decided in the Co-ordination Centre that an appeal should be made to illegal immigrants. The 'illegals' were guaranteed that there would be no negative repercussions if they reported missing acquaintances as well as their own losses. It had been agreed in the Co-ordination Centre that aid and assistance in relation to the air crash would be given to all those in need, irrespective of their legal status. A central telephone number was announced and a contact point was established in the Bijlmer Sports Centre.

On Tuesday morning, those in charge in the Co-ordination Centre decided to accelerate the recovery operations. A timetable was set up whereby they had to be completed by Friday 9 October.

Continued uncertainty

The speed-up of recovery operations soon reduced the volume of rubble around the devastated buildings. Very few bodies were found: the feared number of 250 did not materialise. By the time rescue workers reached the ground floor of the collapsed buildings, on Wednesday afternoon, the remains of 34 victims had been recovered. There was still a significant difference between the actual number of bodies found and the estimated 250 fatalities.

As from Wednesday 7 October, the original estimate of the death toll began to come under scrutiny. The pressure increased further when national authorities started to make enquiries about the death toll. Several embassies had already sought information from both the national government in The Hague and the Amsterdam Co-ordination Centre. Representatives of the Netherlands Antilles urged the Mayor to provide information without delay.

However, the Amsterdam authorities were still not certain about the number of dead. Confusion was compounded by the high number of missing persons reported to the authorities. By Tuesday evening the list of missing persons added up to a total of 1,588 names. The Co-ordination Centre kept these figures confidential, not least because it realised they were out of all proportion to reality.

Verifying the names on the missing persons list was no easy task. For example, the name of the first victim to be identified did not even appear on the list. On Wednesday 7 October a team of 50 detectives set out to clarify the number of missing persons. Meanwhile the RIT was endeavouring to identify the remains in the mortuary, which had been established in a hangar in a remote part of Schiphol airport, with the assistance of Red Cross workers.

On Wednesday the Co-ordination Centre decided that provisional lists of dead and missing persons would be released on the afternoon of Friday 9 October. The time at which the lists were to be released in the Bijlmer Sports Centre was called 'X-Hour'.

Arrival of Lockerbie experts

The recovery crews completed their work at the disaster site on the afternoon of Thursday 8 October, well before the set deadline. The search for the dead came to an end: in total, the remains of 43 people had been found. At the press conference that same day, Mayor Van Thijn said that the air crash had probably cost less than the previously estimated 250 lives. However, the Co-ordination Centre took into consideration that some bodies might have been totally vaporised by the intense heat of the fire. Hence, it might never really be possible to establish the precise death toll.

In the Co-ordination Centre the need for outside expertise was increasingly felt. On Thursday 8 October, two British experts arrived in Amsterdam; both had been involved in the Lockerbie air disaster in 1988. After visiting the Bijlmer site, one of them, pathologist Professor Busuttil, reported that 10% of the dead could have been vaporised.

Material and psycho-social aid

Further help for the survivors got underway during the course of the first week. On Wednesday 7 October the first 10 families were moved into their new homes. The city Social Services provided partial benefit payments for the victims: in the first week after the disaster, 186 people received one-off benefit payments of G 376 to cover basic cost-of-living expenses and, in the same week, some G 550,000 was reserved for the furnishing of homes allocated to victims. The Committee of the European Communities had already allocated ECU 1 million for emergency aid, on Monday 5 October. El Al Israel Airlines also established an emergency fund for immediate relief. El Al spokespeople assured the victims that this fund would not be linked to the settlement of any possible future claims for damages against the airline.

Parallel with this emergency help, on Tuesday 6 October various bodies – under the overall responsibility of Alderman Wildekamp – started to plan for longer-term psycho-social aid. Later that week, the Ministry of Health, Welfare and Culture allocated funds for psycho-social aid for the survivors of the Bijlmer disaster.

Towards the end of the first week, a fraudproblem arose in the form of so-called 'free riders'. These were pseudo-victims attempting to benefit from city provisions and facilities like overnight accommodation, food, cash grants and homes. Suspicions that the number of 'free riders' was growing were reinforced by the rising number of day-time shelter 'clients' at the Bijlmer Sports Centre from Wednesday 7 October onwards.

Sports Centre entrance passes

In order to stem the inflow of non-victims, the Co-ordination Centre decided to introduce a system of entrance passes. Staff duly re-registered the victims, their families and friends, as well as relief workers, and issued them with entrance passes. Some confusion followed later when it became apparent that the city Department of Social Services had set up their own system of entrance passes. In addition, the long queues to get one's pass caused annoyance among those involved.

As the week wore on, the mood in the Sports Centre began to change. At times the atmosphere turned grim. Some people who had not been recognised as victims (and hence were not entitled to relief provisions) became aggressive towards relief workers.

'X-Hour': the publication of the victims' names

In the course of the first week, police enquiries had substantially reduced the number of persons listed as missing. The RIT had successfully put names to several remains. Although identification of the bodies, as well as missing persons enquiries, were still proceeding at full steam, at 18.00 on Friday 9 October ('X-Hour'), the authorities released three lists comprising a total of 120 names. List 1 showed identified fatalities registered to date. List 2 contained the names of missing persons who it was deemed probable had been in or near the area of the flats when the aircraft crashed. List 3 comprised the names of all missing persons from outside the disaster area. Over the weekend the lists were updated and re-released several times.

In the days preceding, much time and effort had been devoted in the Co-ordination Centre to organising the publication of the lists. Mayor Van Thijn continually stressed that the announcements should be as clear and concise as possible. Shortly before 'X-Hour', the Centre's hall started to fill up with family, friends and aid workers. Distribution started in the Bijlmer Sports Centre; only then were lists given to the press. Contingency plans had been drawn up for the possibly widespread traumatic effect of the lists, and plenty of relief workers were in position to offer support. The expected emotional scenes were largely absent.

Reducing relief?

Towards the end of the first week after the crash, the various parties started to consider the cutback of relief provisions at the Bijlmer Sports Centre. The Housing and the Social Services Departments, along with the Nieuw Amsterdam Housing Co-operative, were involved in this discussion. In the opinion of their staffs, the work needing to be done in the sports hall was as good as completed. Moreover, these organisations concluded that – given the rising tension in and around the Centre – it made sense for the city services to withdraw from the Bijlmer Sports Centre. Assistance and support could be continued from the organisations' respective crisis centres which had been established earlier that week.

Discussions also covered the overnight care provisions. Although allocation of new homes was proceeding apace, in the very short term those in need could not be provided with a permanent roof over their heads. Moreover, the navy barracks and other emergency locations could not be used to bridge the gap for this entire period; hence the search began for alternative overnight accommodation. On Friday 9 October, the Antillean authorities took the unexpected step of transferring some 70 members of their community from the Bijlmer Sports Centre to the Park Lake Hotel in Amsterdam. The city authorities – caught by surprise – subsequently felt forced to move the remaining group of victims to a hotel, rather than to another sports centre: they could not allow a discrepancy to develop between the relief provided to Antillean and to other victims. The remaining group of survivors were moved to the Novotel (in Amsterdam) on Monday 12 October. By Monday evening, a total of 197 people had been accommodated in the Novotel.

The memorial service

The Co-ordination Centre had considered the organisation of a multi-cultural memorial service and funeral for victims quite early in the week. A funeral plan was drafted, including the organisation of a collective burial for any unidentified casualties. Preparations were made for a memorial service one week after the crash.

The ceremony to commemorate those who died in the Bijlmer disaster was held on Sunday 11 October. The entire event was televised live in the Netherlands, Surinam and the Netherlands Antilles. That morning, an estimated 15,000 people moved in silent procession from the Bijlmer Sports Centre to the scene of the tragedy. In a short address, Mayor Van Thijn called it 'possibly the hardest walk in our lives'.[10] The many thousands of mourners left behind a sea of flowers at the disaster site. Musicians played from stages erected along the route. The rich cultural mix made it a colourful procession.

Later that afternoon, a memorial service was held in the RAI exhibition complex, attended by more than 12,000 people, among them Princess (formerly Queen) Juliana and Crown Prince Willem-Alexander. Speakers included Premier Lubbers, the Antillean Premier, Mrs Liberia-Peters, Ambassador Ntim Abankwa of Ghana and President Venetiaan of Surinam. Mayor Van Thijn promised long-term help and support for the survivors: 'No one must ever need to say: When I wept, I was alone. This must be our promise'.[11]

End of emergency relief

Relocation of assistance

The original timetable for ending emergency relief had set Monday 12 October as the closing date for relief to survivors in the Bijlmer Sports Centre. The Southeast Neighbourhood Council authorities disagreed. In their view, representatives of the city Social Services and Housing Departments needed to stay on longer in the sports hall. After much discussion, the Amsterdam authorities sided with the Southeast Neighbourhood Council. The Bijlmer Sports Centre was to stay open another week, as a rendezvous and information centre for people affected by the disaster. The Regional Psychiatric Aid Agency (RIAGG), church relief workers and the city Health Department were to be represented at the sports hall during opening hours. The Housing and Social Services Departments were to be in position there for two hours every day.

A definitive list of fatalities

In the second week, police were still receiving sporadic reports on missing persons. In addition to checking out these new reports, the police enquiry team actively looked for missing persons who might not have been reported. In this process a large number of administrative records were used. The enquiries did not produce any major changes in the list of missing persons. Growth was most notable in list 1 (identified fatalities), in line with the success of the RIT in identifying virtually all the human remains recovered. By Friday 16 October, 35 bodies had been identified. The last remains were more difficult to identify: relatives of missing persons were asked to undergo DNA tests, whereby the certain identity of the remains could be established. All the remains were eventually identified,[12] the final toll of the Bijlmer air disaster being 43.[13]

Relief, care and tension in the Novotel and the Park Lake Hotel

The Social Services Department had a tough time managing the floors rented in the Novotel. The group of Antilleans in the Park Lake Hotel was looked after by personnel from FORSA, a support organisation for the Caribbean community in the Netherlands. The relief organisations were unsure exactly how many victims were staying in the hotels. It was suspected that several 'guests' in both hotels were not entitled to relief and support; this feeling was reinforced when the number of 'guests' went up during the week. It appeared that 'free riders' had found their way into both hotels. In order to clear these pseudo-victims from the hotels, it was necessary to conduct an identity check on the hotel populations.

Victims: further screening

The issue of separating 'real' and 'pseudo' victims was discussed in the Coordination Centre on Wednesday 14 October. A working group was entrusted with comparing the data of the various relief agencies.[14] The idea was that integration of records and lists of those affected would give an overall list of the 'real' victims. Alderman Wildekamp had overall management responsibility for the compilation of this 'master list', which would mainly relate to residents of the flats affected. Another route for inclusion in the list was via medical evidence of psychological or physical problems caused by the air disaster; this would also entitle people from outside the disaster area to assistance from the city authorities.

The next step was to screen the people staying in both hotels, though it was accepted that some pseudo-victims might slip through the net. Social Services staff began the screening process on Friday 16 October, using the provisional 'master list' of victims. Anyone not on this 'master list' would be removed from the hotels at short notice.

In the Novotel, 49 people were unable to show that they were resident in the area at the time of the crash; they finally left the Novotel on Saturday 31 October, without much fuss. In the Park Lake Hotel matters went less smoothly. On Thursday 15 October, Alderman Wildekamp had reached agreements whereby FORSA staff would assist Social Services teams in the personal interviews with the Antilleans in the hotel. But, on Friday morning, it transpired that FORSA was not prepared to co-operate with these interviews. Social Services staff reported mounting aggression in the Park Lake Hotel. However, following intensive consultation between the parties involved in providing relief in the hotel, the group of Antilleans staying there were, after all, screened on 19 October.[15]

The departure of the 'non-recognised' group from the Park Lake Hotel was not without problems; some of them – under threat of imminent removal – issued a writ against the city authorities. They demanded that the criteria for victim status should be made public and that the court should judge on the correctness of the procedure used. The court ruled in favour of the city.

Residence permits for illegal immigrants

On Tuesday 13 October, the authorities' earlier promise that no differentiation would be made in the assistance given to victims, irrespective of their immigration status, was put to the test. Due to faulty registration of their personal details, six Ghanaians who had reported in after the disaster were referred across to the Police Immigration Department. There, the Ghanaians were ordered to report to the police on a regular basis, and threatened with deportation. Once the mistake was discovered, Chief of Police Nordholt ordered their files to be destroyed. But the authorities' supposedly broken promise provoked a public uproar. The very next daily press conference was interrupted by a bomb scare – which the police spokesman linked to the uproar surrounding the illegal Ghanaians being sent to the Immigration Department.

On Wednesday 14 October, Chief of Police Nordholt consulted with State Secretary for Justice Kosto[16] on the possibility of granting residence permits to illegal immigrants among the victims of the air disaster. After their meeting, the State Secretary and Amsterdam's senior police officer gave a press conference. Kosto said that he was prepared to look favourably at granting residence permits to the illegal immigrants involved, on a case-by-case basis; these would be illegal immigrants who had been offered a home as part of the city of Amsterdam's relief efforts. It was stressed that this *ad hoc* legalisation did not imply a general pardon.

Kosto's statement created the impression among illegal immigrants that any of them who could demonstrate that he or she had been affected would almost certainly be legalised. At the same press conference, Nordholt had said that a list of persons who might be entitled to a permanent residence permit would be compiled under the authority of Alderman Wildekamp. Secretary Kosto would then use this list as a basis for his final decision on legalisation of the people concerned. This list became known in the press as the 'Kosto list'.

The Kosto list

The Amsterdam city authorities had deliberately made the proviso that they were not aware of every single victim. Missing people might have temporarily gone into hiding in private homes, but there might also be some illegal immigrants who had not yet reported in. The 'illegal victims' who had not reported in might still qualify for the Kosto list. On Friday 16 October it was decided in the Co-ordination Centre to set a time limit: people affected by the air disaster were given until 18.00 on Saturday 17 October to register with the Population Register's contact desk in the Bijlmer Sports Centre.

State Secretary Kosto's agreement that illegal immigrants who had been affected would be considered for legalisation was meant to give this group the confidence to report in. The Population Register officials would then judge whether the persons registered had actually lived in the disaster area. After investigation, the illegals among those affected would then be put on the Kosto list and hence presented for legalisation.

On Saturday 17 October hundreds of illegals came to the Bijlmer Sports Centre for registration. This massive influx meant that not everyone could be processed. These people were then asked to report for registration to the Population Register on Monday 19 October. The rumour that one could be registered at the Population Register on Monday spread fast among groups of illegal immigrants. However, the normal procedures for obtaining a residence permit applied in the case of this group of illegals. When registration of the illegals had been completed, almost 1,800 people had requested legalisation at the Population Register. On Thursday 22 October the Co-ordination Centre expected that only some 100 illegals would be considered for a residence permit.

Aftermath

Long-term relief

On Tuesday 27 October Amsterdam's governing body, the Board of Mayor and Aldermen, approved the after-care plan for victims of the Bijlmer disaster. Alderman Wildekamp had presented the long-term relief plan in the Co-ordination Centre, on the morning of Monday 12 October. The idea was that, as far as possible, relief and support would be provided in Southeast Amsterdam. Specific contributions from neighbourhood bodies and immigrant groupings had been built into the plan, alongside expert input.

Two locations would serve as centres for after-care activities up to mid-November. After closure of the Bijlmer Sports Centre on 19 October, an advice and information bureau was established in a nearby shopping centre. Legal advice was also available for a period of two weeks. From 19 October onwards an ecumenical centre served as a walk-in facility for those directly involved, and for local people. Care and support resources were in particularly heavy demand in the early period.

Illegal immigrant issue

Once the victims' after-care plan had been established, on Tuesday 27 October Mayor Van Thijn rescinded the disaster declaration. But there were still people in the Park Lake Hotel and the Novotel, some of them waiting for new homes to be made ready. It would be some months before the last Bijlmer disaster victims left the two hotels: the last group checked out on 22 December.

The Population Register was also stretched to its limits in the wake of the Bijlmer disaster. This city department had to check the details of the almost 1,800 illegals who had registered there as being affected by the plane crash. In the months following the crash, the Director of the Population Register made recommendations to the Mayor on the applications received. A list of 91 names was submitted in April 1993 to State Secretary Kosto; of these, 55 have been granted residence permits.

The Bijlmer air disaster put the spotlight once again on the issue of illegal immigrants in the Netherlands. The long queue of illegals waiting to get into the doors of the Population Register had spurred heavy press coverage. The scale of the illegal immigration problem in the Netherlands, and how best to approach it, became the subject of public debate, the general tenor being that a tougher approach was needed, in Amsterdam and elsewhere, with greater scope for checking immigration status and stronger sanctions on rogue employers.

Air traffic over Amsterdam: renewed discussion

On Monday 5 October the Director-General of the Netherlands Civil Aviation Authority (Rijksluchtvaartdienst) announced an immediate ban on all flights to and from Schiphol airport over the disaster area in Southeast Amsterdam. The main reason for this decision was psychological: the sight and sound of aircraft over the Bijlmer so soon after the crash could be emotionally damaging.

On Tuesday 20 October, take-offs and landings were resumed on Schiphol airport's Buitenveldert runway and flights started again over the disaster site. At the same time, air traffic over Southeast Amsterdam was built up in gradual stages, from small to heavier aircraft. By Saturday 24 October normal flights had resumed over the Bijlmer, 12 of whose residents issued a writ aimed at banning the resumption of flights over their neighbourhood. They demanded that a new ban be instituted for a period of six months; the court rejected this request on 20 November.

The Bijlmer disaster prompted renewed discussion of existing plans to expand Schiphol airport. Particular attention was now paid to safety aspects, which had hitherto played a subordinate role in the expansion plans. On 14 October the Board of Mayor and Aldermen wrote a letter to the City Council to the effect that a reconsideration period was essential with regard to the Schiphol expansion plans. During this time, primary attention should go to risks entailed by increased air traffic over the city. The Board decided that no new home construction projects for the area under the approach and departure flight paths would be considered prior to the summer of 1993. Since its formation on 18 November 1992, the Bijlmer Air Traffic Working Party has also been reviewing expansion plans for Schiphol, the country's main international airport. As a direct consequence of the air crash, work on expanding Schiphol was halted for six months. Eventually, however, the safety concerns raised by the Bijlmer disaster began to wane and the Schiphol expansion plans went ahead undiminished.

Surveying the community impact

In mid-March 1993 (five months after the crash) we commissioned a telephone survey of 363 Amsterdam residents.[17]

Death toll

The respondents were asked if they knew how many people had died in the disaster; their answers showed a tendency to over- rather than underestimate the figure. Eighteen per cent of those polled put the figure far too high (more than 70 dead). Far and away the largest group (44%) thought that between 55 and 70 had died; and 28% gave a figure of between 30 and 50 dead. Although – given the timing – not too much weight should be attached to these figures, they do indicate some signs of the so-called 'primacy effect': the picture of the first few days after the crash stuck in the minds of many Amsterdam citizens, and later downward adjustments in the death toll often failed to register.

Involvement

Asked whether family, friends or acquaintances lived in or near the area affected, 41% of respondents – who did not themselves live in Amsterdam Southeast – answered in the affirmative. This figure gives some indication of the widespread involvement of Amsterdam citizens with this disaster in the Bijlmer. Apart from the actual list of standard questions, the poll also gave respondents the opportunity to make their own remarks about the disaster. About one in three did so, sometimes at length – proof that events had become enshrined deeply in Amsterdam's collective memory.

Visibility of organisations

Asked which city services were involved in the disaster, respondents mainly cited the organisations and services that played a key role in the acute phase (Table 1.1). Understandably enough, most respondents emphasised the first few minutes and hours of the disaster. Subsequent events and the key organisations in these phases were much less likely to be associated with the disaster.

Most frequently mentioned were the police and fire brigade (both 78%). These were followed by the Health and Medical Services Department, the Central Ambulance Station (respectively 44 and 27%). Organisations like the city's Housing Department and Population Register, which fulfilled complex and pivotal roles later in the process, had relatively fewer mentions (respectively 11 and 5%). As is always the case with major civil catastrophes, the Bijlmer disaster was overwhelmingly associated with the police, fire and medical services – the people in uniform.

TABLE 1.1 *The 12 most frequently mentioned organisations.*

Organisation	% respondents
Police	78
Fire brigade	78
City Health and Medical Department	44
Central Ambulance Control	27
City Social Services	21
Police Disaster Identification Team (RIT)	15
City Housing Department	11
Regional Psychiatric Aid Agency (RIAGG)	10
Salvation Army	7
Red Cross	6
City Population Register	5
Southeast Neighbourhood Council	4

Notes

1 This study does not go into the media controversy surrounding the role of the Civil Aviation Authority (Rijksluchtvaartdienst) in investigations of the cause of the crash of the El Al Boeing 747.
2 *Het Algemeen Dagblad*, 6 October 1992.
3 Figures provided by the Nieuw Amsterdam Housing Co-operative.
4 *Het Parool*, 17 October 1992.
5 Alongside the Mayor and Chief of Police, who both arrived at City Hall at 19.10, the officials who took up position in the Co-ordination Centre were Fire Chief Ernst; Head of the City Information Department, Mrs Van Oostveen; Deputy Head of the Medical Services Department Coutinho; Director of City Transportation Smit; Deputy Council Secretary Wessels; Alderman for Cultural Affairs Bakker; and Director of City Social Services Van Dijk. The Mayor was further assisted by staff of the General Administrative and Legal Affairs Department, and in particular by staff of the Public Order and Safety Section. Deputy Chief Prosecutor Mijnssen came to the Co-ordination Centre on his own initiative.
6 *Trouw*, 6 October 1992.
7 *Gemeenteblad*, section 2, 1150.
8 It was widely suspected that many illegal immigrants had found their way to the Bijlmer.
9 *Gemeenteblad*, section 2, 1150.
10 *Het Algemeen Dagblad*, 12 October 1992.
11 *Het Parool*, 12 October 1992.
12 The last two bodies were identified in April 1993.
13 In theory it is possible that some fatal casualties in the disaster were never reported missing to the authorities. However, our respondents in the fire service, police, the RIT and the Co-ordination Centre regard this as most unlikely.
14 The working group consisted of representatives of the Southeast Neighbourhood Council, the police, the Population Register, the Social Services and Housing Departments, and City Hall's General Administrative and Legal Affairs section.

15 These consultations were chaired by Deputy Mayor De Grave. Mayor Van Thijn had left for Indonesia on official business on 16 October.

16 The term State Secretary (in Dutch: Staatssecretaris) is used to describe an official who in other countries is known as a Deputy Secretary or Under-Secretary or Junior Minister.

17 For a more detailed report of this public opinion poll, carried out by Intomart, see Appendix 3.

CHAPTER 2

Disaster response

Introduction

The first minutes and hours after a sudden disaster has struck are generally a period of chaos and disorder.[1] Once this initial period has passed, the authorities tend gradually to assume some control over the situation. In other words, initial chaos is followed by attempts to reimpose some degree of order. The crux of counter-disaster preparations lies in keeping the duration of the chaos period to a minimum; gaining time is an important function of preparation. Meaningful preparation should also enhance quality: not only can the authorities gain control of the disaster situation more quickly, but they can also ensure that better intelligence is available and that the appropriate manpower and equipment are deployed to this end.[2]

This chapter describes counter-disaster measures after the Amsterdam air crash. It focuses primarily on the city authorities' initial responses, and looks at how they dealt with the organisational and communication challenges posed by the event. Following a more detailed description of the first few hours of the disaster period, we shall discuss various aspects of the initial response by comparing the Amsterdam situation with some of the more general findings of previous research into quick-response mobilisation and disaster response.

Mobilisation: the first few hours

Several city agencies had already received the alarm prior to the aircraft going down. At the very moment of impact, the police, fire brigade and ambulance services all received calls from eye-witnesses as well as from Schiphol airport authorities reporting an aircraft in difficulties. And immediately after impact, the emergency services were swamped with calls about the crash in the Bijlmer.

Parallel to this, some key officials had already been alerted via more or less informal channels. A City Hall concierge living in the Bijlmer saw the aircraft crash and called his duty colleagues right away; in turn, they followed standing orders and informed a whole range of authorities. An official of the North Holland provincial administration caught reports that an aircraft was in trouble on the police radio wavelength, and promptly called the provincial duty officer. A Director-General of the Ministry of the Interior was driving along when he saw the aircraft coming down and promptly called his colleague, the Director-General for Public Order and Safety, whereupon the alarm went out to various people at the Ministry.

After these initial exchanges both the informal channels and the media ensured that, within 30 minutes, dozens of fire-fighters, police, doctors, nurses and other relief workers were aware that an aircraft had crashed on the Bijlmer. A large number of police officers immediately rushed towards the Bijlmer on their own initiative. Without hesitation dozens of individuals and organisations volunteered expertise, means and input. Many donors made their way to hospital to give blood. Inevitably also, these initial reports and communications quickly resulted in a large number of journalists, photographers and sightseers flooding towards the disaster area. Within a short period of time, the number of people seeking to make their way to the site of the crash had grown so large that police closed several roads leading into the Bijlmermeer area. This was the classic 'mass convergence' that almost always occurs when a major disaster strikes in a populated area.

At the scene

Within minutes, fire, police, medical and ambulance personnel were on their way to the scene of the accident. A pump appliance and turntable ladder were scrambled from the local fire station, and the alarm was sent to other stations. Although the first reports mentioned an air crash in Diemen, it was soon clear that the apartment blocks 'Kruitberg' and 'Groeneveen' in the Bijlmer had been hit. At 18.47 hours a fire-fighter radioed to the emergency control room:

> 'We're still the only ones here and you've never seen anything like it in your life. One won't be enough, make sure you send a dozen pump appliances.
> *Everything we've got is on the way, pump appliances, foam tenders, the whole shebang's coming. I understand it's a very large fire, over.*
> Yeah, it's the biggest I've ever seen.
> *Hold on, what sort of plane is it? Is it a Boeing, over?*
> You can't tell any more. There are 40 ... 50 to 60 flats blazing away like hell. Yeah, there's something lying there like a plane, that's it, over.
> *OK, we're doing everything we can. You'll get help as soon as possible, over and out'.*[3]

In addition to the fire brigade, the police and medical services directed a large number of vehicles to the scene. The emergency services arrived on both sides of the flats. The police and fire brigade control centres had already been in contact about the best available route and closing off the roads. By 19.00, all the emergency services were aware that a freight-carrying aircraft was involved.

The first fire-fighters on the spot turned their hoses on the various ground fires. As quickly as possible, the police tried to seal off the disaster site, forming two rings – the first to seal off the immediate area of the blazing flats and the second to shield the part of the neighbourhood in which the flats were situated. Ambulances took the first of the injured to the Onze Lieve Vrouwe Hospital. Within approximately 40 minutes after the crash the two large university hospitals in Amsterdam, the Academic Medical Centre (AMC) and the Academic Hospital of the Free University, reported that they had activated their disaster plans.

Within the first few hours the fire-fighters were able to extinguish the separate fires. Shortly after 21.00 it was announced that the fire was under control and all the fires were put out in the course of the night. Meanwhile, 11 people had been taken to hospital by ambulance. Taxis and private cars had taken a further 25 light casualties to the AMC. By then, the police were busy warding off sightseers and overly intrusive photographers.

In the Co-ordination Centre

The people with positions in the city's disaster staff or one of the city agencies' action centres were all notified immediately. The Amsterdam Co-ordination Centre in the basement of City Hall was fully operational and at work within one hour.

Despite this efficiency in getting organised, it turned out to be far more difficult and time-consuming to establish direct contact with the disaster site. That notwithstanding, the people in the Co-ordination Centre tried to garner information on events in and around the disaster site via various ways and means. Confusing reports caused a series of misunderstandings that first evening. For quite some time it was believed that eight – later as many as 12 – bodies had been recovered, a spokesperson for the AMC hospital having given this information to radio and TV reporters. In fact, at this point not a single victim's body had been found.

During the first evening the most important activities were centred on the disaster site, and it was there and then that important operational decisions were taken. As the evening wore on, the Co-ordination Centre obtained a better picture of the situation and was able to exert more control over the operation, as heads of services established direct contact with their units in the field and attempted to command them from City Hall. A second factor enhancing the impact of the Co-ordination Centre was the centralisation of the city authorities' information activities at City Hall.

Bottle-necks

There were several bottle-necks during mobilisation on the first evening, for instance the problematic location and accessibility of the disaster site. Given the idiosyncratic street name pattern of the area, the flats were hard to find and it proved difficult for the emergency services and equipment to reach them. Since the telephone lines around the flats were out of order, it took hours before proper links could be established with the Co-ordination Centre in City Hall. So for an extended period the Co-ordination Centre had only sketchy, often unverified reports to work with.

Growing numbers of relief workers arrived at the disaster site, creating an unusable surplus of manpower. Groups of relief workers from outside Amsterdam also arrived at the Bijlmer on their own initiative. And there was constant traffic with city officials and the heads of the various emergency services coming to check on developments on the spot. This prompted one city Health and Medical Services worker to say:

> The problem with disasters is that so many people want to get involved. The Health Inspector arrived at the Central Ambulance Station just over an hour after the first alarm had been raised, followed by the Director-General for Public Health and the Chief Public Health Inspector. Personnel doing the work saw the presence of these VIPS as an extra burden. Some people felt that it hindered their performance.[4]

Another bottle-neck was that, initially, the relief and assistance was provided to survivors in many different locations. Indeed, for some time the Co-ordination Centre was uncertain precisely which buildings were being used as relief posts. During the course of the evening it was decided to centralise relief on a single location, i.e. the Bijlmer Sports Centre (see Chapter 4).

The large number of sightseers and press posed a problem for the police. During the evening, two units of the police mobile riot forces were deployed to seal off the disaster site from sightseers. The fact that in that process no clear distinction was made between inquisitive civilians and, in some cases, pushy journalists caused a significant amount of irritation back and forth. Lastly, matters were complicated by the inability to establish contact with the authorities of the Southeast Neighbourhood Council on the first evening. Further, although the Neighbourhood Council did not have a presence in the city's Co-ordination

Centre during this period, they were carrying out a whole range of activities on the spot.

Intermezzo: the institutional context of disaster management in Amsterdam

The Bijlmer air crash was in the capital of the Netherlands. A historic city embedded in an urban region of close to one million people, Amsterdam has always been a complex place to govern. It has a long tradition of resistance to authority and its history is marked by demonstrations and disorder. Most recently, the early part of the 1980s saw an increasingly bitter and violent confrontation between local authorities and police on the one hand, and a well-organised and determined movement of urban squatters, mostly young people from out of town looking for affordable housing and frustrated by the coincidence of significant housing shortages in the old part of town and widespread speculation practices by real estate owners resulting in numerous empty, unused premises in that same area. Occupying these buildings and turning them into dwellings via a comprehensive alternative distribution system brought the squatters into direct confrontation with the police, who in most cases were obliged by court orders to evict them, and with local authorities, whom they accused of complacency in the face of an intolerable situation. The result was a series of street clashes, involving large crowds and massive riot police operations.

For these reasons, the Amsterdam authorities and emergency services are no strangers to urban crisis management. At the same time, the city has seen relatively few major civil disasters in the last few decades, the most significant being an industrial explosion in 1971 and a major hotel fire in 1977. For all the attention given to mitigating civil unrest and controlling disorder, the city of Amsterdam had much in common with most other Dutch cities when it came to planning for and experience with major disasters: this was regarded as a relatively low-priority issue – until the disaster happened.

The events of the Bijlmer air crash and the official response to them need to be interpreted from the proper perspective of the broader institutional context of government organisation and disaster planning in the Netherlands. This will now be briefly described.[5]

Governmental system of the Netherlands

The Netherlands, a small country squeezed between Germany and the North Sea, is home to approximately 15 million people and one of the most densely populated countries on earth. It is also one of the most prosperous, whose economy has been boosted by a long trading history, the largest harbour in the world, and its strategic location as gateway to Europe. It is a decentralised unitary state with three levels of government, i.e. central government, 12 provinces and several hundred municipalities. Since the enactment of the 1848 Constitution, local government has been protected from formal interference by central authority. Based on this principle an elaborate system of co-governance between local and central layers of public administration has developed, with the provincial level traditionally being relatively weak.

Each municipality has a Municipal Council, the office of Mayor and the executive Board of Mayor and Aldermen. Being the municipality's supreme body, the Council is elected by universal suffrage within a system of proportional representation. In day-to-day practice, the Board of Mayor and Aldermen, which often consists of full-time professional administrators, plays a leading role, with the Council generally limiting itself to legislative oversight and control of the executive Board. The Aldermen are elected from the Council, while the Mayor is appointed by the central government. Historically, Mayors were supposed to act as the eyes and ears of the central authorities, and to enforce central rule if necessary. Nowadays, Mayors have become purely local officials who will not shy away from tough negotiations with their superiors at the Ministry of the Interior in The Hague to defend local interests. In fact, as Chairman of the Council and the Board, the Mayor stands out as the municipality's 'first citizen'. At the same time, the Mayor does carry out some tasks emanating directly from formal responsibilities imposed by central government.

Municipalities have both autonomous and co-management tasks. The most important tasks delegated to or implemented by the local government are the maintenance of public order and fire protection, the upkeep of roads, and several tasks related to public transportation, economic policy, primary education, culture, social welfare and health care.

The structure of the government of Amsterdam conforms to this general model with one important exception. As a late result of the democratisation movement, the city in 1990 established a number of so-called Neighbourhood Councils, which essentially are sub-governments within the city domain, each covering a well-defined part of town. These Neighbourhood Councils are also elected through direct and general elections, and they too have an executive Board of professional administrators as well as an executive bureaucracy. They, in turn, have been charged with several tasks and responsibilities delegated to them by the city, but can also undertake initiatives on their own accord. Not long after they were established and despite a rather meagre turnout at the polls and significant administrative problems in getting their organisation set up, some Neighbourhood Councils quickly developed into kinds of countervailing powers against 'the city'. Ironically, at the same time major urban centres such as Amsterdam are being caught up in a process of regionalisation, with, for example, the formation of administrative structures at a fourth, intermediary level of government. In fact, some executive agencies crucial to disaster management, such as the fire brigade and the police, have in recent years switched nationwide from a local to a regional mode of organisation – with the precise boundaries of the regions differing markedly across organisations. These parallel developments have complicated the administrative structure of cities like Amsterdam and Rotterdam.

Emergency planning in the Netherlands

In 1985, the Dutch Parliament passed a new Disaster Act. This Act defines a disaster as an event that endangers the life and health of a large number of people, or causes severe harm to material interests, and which requires co-ordinated efforts from various fields of expertise. This conceptualisation of disaster reflects the prevailing principles of disaster management in the Netherlands. A very general notion of the social and economic disruption of the community is combined with an explicit demand for governmental activity and for co-ordination between governmental agencies.

The Disaster Act has several provisions with regard to disaster planning procedures. The Board of Mayor and Aldermen is supposed to draw up a mandatory disaster plan comprising a list of the agencies that may have to take part in emergency activities. The Municipal Council has to approve this plan. The Mayor, who has the primary responsibility for disaster response after having issued a formal disaster declaration, is obliged to make a special operational plan, including a response scenario for possible calamities in specific disaster-prone sites in the local community, such as petro-chemical plants, high-risk traffic routes or industrial zones. These operational plans are classified.

It should be stressed that the division of tasks and responsibilities that occurs in a particular disaster may adapt itself to the evolving circumstances of the event in question. In the preparedness phase of emergency management the municipal authorities may be the key actors, which betrays the essentially decentralised approach envisaged in the Act. But when the situation becomes more serious and/or transcends the boundaries of one municipality, provincial or even national authorities (especially the Ministry of the Interior) may decide to co-ordinate or otherwise intervene.

Many different organisations and institutions are involved in emergency planning – all in different ways and to a different extent. The most important operational services involved are the local and regional fire brigades. As the Disaster Act dictates, the Fire Chief has the primary responsibility for the on-site co-ordination of local disaster responses. Every municipal administration in The Netherlands employs one or more civil servants involved in disaster planning activities, but their impact has in most cases been limited. In the bigger cities, however, particularly Amsterdam and Rotterdam, which both have extensive industrial and harbour areas housing a significant number of high-risk installations, municipal counter-disaster officials have taken a more active role in setting up the necessary infrastructure and organising exercises.

Counter-disaster operations: from planning to reality

In the Netherlands as well as in some other European countries, the media and the authorities always attach great importance to the existence and implementation of a disaster plan. Indeed, whenever a major disaster occurs some of the media appear totally preoccupied with the whole issue of these plans. For some of the officials involved in the handling of the Bijlmer disaster, the plan appeared to play an overly dominant role. Within two hours of the crash, the Minister of Transport and Public Works stated that the Amsterdam disaster plan had worked well. But some weeks later, the Amsterdam Fire Chief caused not a little surprise when he said that 'there was never any need to take the disaster plan off the shelf'. Even more notably, he added that, 'It was a great benefit that the plan wasn't used'.[6]

The one statement appears to contradict the other. In turn, this raises the question of how, in terms of disaster planning, best to analyse and evaluate the city's performance during the first few hours of the disaster. Drawing upon the international literature, four main themes for such an assessment have been selected.

Raising the alarm and issuing alerts: basic notions

In disaster situations, much time is always used in sounding the alarm and alerting the important organisations, services and individuals. Not infrequently, after several hours have passed, it transpires that important persons have not been alerted, or that there is a sudden need to reach a given organisation – which had not been included in the plans.[7]

The availability of a disaster plan is a necessary condition for achieving a fast and efficient alert, rather than a sufficient guarantee that this will indeed happen. For this to occur, it is essential that the planned procedures are regularly tested and practised in a range of different situations. It is also a fact that an organisation that regularly deals with extraordinary circumstances requiring speedy mobilisation will often be able to mount a fast response. Quick mobilisation becomes a matter of routine: people become familiar with the strong and weak links in communications and with network agreements. Where needed they can be adjusted. The individuals and organisations co-operate better by virtue of experience.[8]

Alarms and alerts: the Amsterdam experience

The alarms and alerts proceeded relatively effectively and efficiently in Amsterdam. Within a short period a large force of fire-fighters and police with equipment were deployed at the scene. And the medical chain – ambulances, trauma teams, first aid provisions, casualty distribution, hospitals – were on-line to treat and transport large numbers of casualties.

The operational services were not the only ones to react quickly. The city officials designated by the disaster plan to support and advise the Mayor in the Co-ordination Centre were also present and active within the space of a few minutes, partly benefitting from the fact that some of them lived close to City Hall. The Co-ordination Centre in the City Hall basement soon filled up with designated representatives of various city organisations and services. The extensive infrastructural facilities installed below ground level proved to be anything but a luxury.

Good planning appears to have contributed to Amsterdam's fast mobilisation at such an inauspicious time as early Sunday evening. The black-on-white alert schedules used by the city services and the co-ordination centres had been tested in practice before and worked well again this time around.

Routinisation is also a factor in a city like Amsterdam. Major incidents like the Bijlmer disaster are rare, but the city administration regularly has to deal with extraordinary situations demanding a fast response and mobilisation of services and personnel. Examples of this are sudden, unexpected large-scale disturbances of public order; alongside decision making at city administration level, these may also require supply operations by city services like cleansing and transportation. As mentioned above, this type of disturbance was particularly common in the 1980s.[9] Events involving mass crowds, such as major sports events, demonstrations and public celebrations, regularly require speedy, large-scale deployment of city services.

The only truly serious problem arising in the first few hours after the alert was that the Southeast Neighbourhood Council could not be contacted; their representatives were not aware that they should have been present at City Hall in a disaster situation. One likely reason for this has been referred to earlier: that, as a relatively new phenomenon in the city government structure (Neighbourhood Councils were established in 1990), they were as yet less than fully slotted into the city's disaster plan.

A second problem did not affect the speed and quality of mobilisation; rather, it was the unintended outcome of this very efficiency. By the end of the first evening, there turned out to be a relatively small number of wounded people. Not only was this greeted by a feeling of relief, but there were indications of surprise, anti-climax and frustration among the medical aid workers who had been mobi-

hospital beds had been made ready; the specialists had been 'bleeped'; the nursing staff were present in large numbers. But the expected stream of wounded never arrived. The crash of the aircraft had injured relatively few people. People whose flats received a direct hit were probably killed instantaneously. Most of their neighbours were able to escape from the resulting fire, in good time and with only limited injuries.

These facts were not immediately known in the first few hours following impact. It proved difficult for the medical teams fully to digest this extraordinary course of events. After a rushed mobilisation and several hours standing by at full alert, it was 'back to business as usual', without actually having had to do much. While other organisations were tested to the full, for the medical services the Bijlmer disaster was in many ways nothing more than a large-scale exercise for alert and deployment procedures.

Rapid co-ordination

Basic notions. In the early hours of a calamity situation it is never easy to establish speedy and meaningful co-ordination between the various organisations and services and between the operational and policy levels (see also Chapter 6).[10] Improvisation is the most notable characteristic in dealing with the initial phase of disaster response. Contingency planning should aim at establishing this co-ordination at the earliest possible stage and, as far as is feasible, to bring a little structure to all the improvisation that is taking place.

Different disasters call for deployment of a very different mix of organisations. Obviously, the fire, police and medical services – often collectively designated as the 'emergency services' – are most likely to be among the crucial organisations in the early stages. But from then on what makes the difference is whether the disaster is an earthquake, an explosion, a radiation leak or an accident with petro-chemicals. The precise involvement of municipal and other organisations will largely be dictated by the specific nature of the incident. Accordingly, issues of co-ordination will differ per type of disaster.[11]

The Amsterdam experience. Right from the start, the operational services most involved had regular contact at the scene. The on-site commanders of the fire brigade, police and medical services positioned their communication and command vehicles close together. Later that first evening, the various service chiefs were in close contact while present in the Co-ordination Centre.

Where co-ordination problems did arise, these concerned relations between the operational services on the spot and their senior executives in the Co-ordination Centre. Initially the Co-ordination Centre had no direct links with the operational commanders at the scene of the disaster. Moreover, the Co-ordination

Centre had minimal insight into the care of victims in the various relief centres. So, for example, that first night there was scarcely any contact between the Bijlmer Sports Centre and the Co-ordination Centre. Hence, central co-ordination in the initial phase was minimal.

Even so, it is fair to ask whether people on the spot perceived this as a real hindrance. It has to be borne in mind that there were two compensatory factors. By Dutch standards, Amsterdam has considerable resources, expertise and – as previously noted – manpower familiar with crisis situations. This applies both to City Hall and the municipal services. Consequently these organisations were able to perform reasonably autonomously. Most co-ordination problems could also be dealt with on the spot. For example, on the first evening a number of police sergeants and inspectors were able to organise first-line relief and assistance in the Bijlmer Sports Centre.

The disaster also provoked a massive influx of volunteers, whether or not affiliated to bodies like the Red Cross and the Salvation Army. Meaningful deployment of these people also imposed demands but, once in position, the volunteers often carried out useful supporting tasks (see Chapter 4). They formed a logistical buffer and enhanced the flexibility of relief efforts.

Uncertainty and communication

Basic notions. Widespread uncertainty is a common factor in the immediate wake of a disaster. It is often hard to say precisely what has happened and what the consequences are. Disbelief is the first reaction when people hear the news: 'It can't be happening. Not here. Not now.'[12] Then the first confirmations start to come in. From then on the authorities and aid organisations are commonly swamped by an often contradictory stream of incoming reports and rumours. In contrast, one also notices lulls in the storm: those remarkable periods when everyone has comprehended that something very serious has happened and a whole range of responses to misfortune has been initiated, but the authorities are getting little or no concrete information on the situation on the spot. This is a very difficult period for them, since they are expected to form a picture of the progress of the disaster response, and of any supplementary measures needed.

It usually takes quite some time before communication links are working well enough for the key organisations to establish operational contacts with one another. In the first phases of disaster, not only will public communications, telephone lines in particular, get swamped, but reporting within and between organisations will also log-jam. This may be the outcome of inadequate equipment, but poor 'radio discipline' among the personnel of the relevant organisations may also be part of the problem.[13]

The Amsterdam experience. In the Bijlmer case, there was indeed widespread uncertainty for the first few hours. Many found it hard to comprehend what had happened. It took quite some time before most communication links were working well, e.g. between the disaster site and the Co-ordination Centre. Some people stationed in the Co-ordination Centre had trouble operating the available communications equipment.

The media played a significant role in the information process. On the one hand they fed the rumour circuit while, on the other, at the very least they provided a tangible picture of the situation to many people – including the decision makers. Local, regional and national radio made a particularly significant contribution here.

The question of the death toll kept the Co-ordination Centre uncertain during the first evening, and in fact throughout the week. On the Sunday evening when the disaster struck the initial situation was so confused that the authorities were unable to announce any estimate of casualties. That first evening the media were speculating about possibly 150 to 200 people having been killed – an estimate obtained from the police, one officer having let slip this figure when talking to a reporter.[14]

Media behaviour and impact

Basic notions. Disasters and other crises are media events par excellence. Media people spare neither cost nor effort in securing immediate on-the-spot pictures and maximum information. This may bring media interests into conflict with those of the authorities and relief workers. Disputes can easily arise at the disaster site in particular. The emergency teams on the spot are trained to perform their tasks as fast and efficiently as possible. The massive influx of reporters may hinder this. Emergency workers on-site see the media primarily as an obstacle and have little appreciation for the fact that press and broadcasting can perform an important service in keeping the public informed. Pushiness by reporters, photographers and camera crews can exasperate relief workers because it shows so little consideration for the predicament and feelings of the victims. It often needs considerable effort at the policy level to secure an on-site situation whereby the media can get on with their job, within reasonable parameters, while not getting in the way of relief operations. At the same time, from the first stages onwards, the media may also pressurise the authorities with awkward questions as to what caused the calamity. Inevitably this leads into questions about responsibility and liability.[15]

The Amsterdam experience. Amsterdam is the very hive of Dutch media. No town or city in the Netherlands has so many journalists, press photographers and other media people among its citizens. This makes it hardly surprising that so many journalists were on the spot so quickly. Most city agencies with a direct or indirect involvement in the disaster were swamped by press people. The media were present in force: at City Hall, at police headquarters, at various fire stations and hospitals, at the Southeast Neighbourhood Council and relief centres. They were anxiously seeking picture opportunities and scarce information. The veritable legion of journalists that had made its way to the scene of the disaster was hampering relief operations there. Amsterdam's high international profile also drew the foreign media: this was a potential major world-class story. CNN even scored a TV scoop in the Netherlands when it broadcast a report of the crash at 19.23.

Intrusive journalists and camera crews were hindering the emergency services in combatting the disaster and helping victims. After police units had cleared unauthorised persons from the immediate area of the flats, the rest of the evening and night saw considerable irritation flare up between police officers and journalists. Indeed, on Tuesday 6 October, the Dutch Journalists' Union submitted a written complaint to city Police Chief Nordholt and Mayor Van Thijn, accusing the police of 'very rough' treatment of journalists.[16]

The pressure on the field commanders and personnel of the emergency services and the authorities mounted quickly. The media spread unverified reports and popular disaster myths about panic and mass looting. In particular, the looting that was alleged to have broken out in the Bijlmer received much press and police attention. In retrospect, all this turned out to be greatly exaggerated: there was plenty of talk of looting (*vide* the many reports of plundering in the police log book) whereas, in fact, there was not one single confirmed incident on the first evening.

The first steps to centralise the city's information to the media, as far as possible, were taken at around 19.30. The Head of City Hall's Information Department took control from her position in the Co-ordination Centre, and met the assembled press twice that evening. But it soon became clear that journalists wanted to see the people in authority. For that reason, Mayor Van Thijn, Fire Chief Ernst and Police Chief Nordholt attended an improvised press conference. City Hall's information and public relations staff dealt with questions from domestic and international reporters throughout the night. The Amsterdam spokespeople were unprepared for the burgeoning and specific demands of the international press corps for English language communications. This represented an additional burden.

During the first few hours there was an unfortunate incident in regard to reports on the number of bodies recovered. City spokespeople confirmed an incorrect report that 12 bodies had been taken to one of the hospitals. This report had previously been released and confirmed in the Co-ordination Centre.

Conclusions

Viewed from the perspective of acute counter-disaster operations, there were several aspects that made the Bijlmer disaster an atypical event. The low number of severely and lightly injured limited the significance of the medical component in relation to the scale of the disaster. Moreover, in this disaster situation, there was no direct link between the speed and efficiency of counter-disaster operations on the one hand, and the scale of the disaster in terms of casualty figures on the other. Those who died, died instantaneously when the aircraft crashed. No matter how fast their mobilisation, there was little the fire, police and other emergency services could have done to limit the number of casualties.

A second atypical aspect of the disaster was that it happened in Amsterdam, the city with the largest police force in the Netherlands, with a large regional fire service, plenty of ambulances and hospital beds, and a significant manpower potential. Moreover, the most senior city officials had adequate experience in dealing with crises. It is unlikely that a smaller municipality would have had the capabilities to cope with a similar calamity on its own.[17] This background is important for any meaningful evaluation of emergency counter-disaster activities after the Bijlmer crash.

An important positive element in acute counter-disaster operations in the Bijlmer situation was the speed with which the city's relief workers, managers and volunteers got into position on hearing the report of the disaster. The various duty units of the emergency services speedily mobilised manpower and went into action. A contributory factor in the fast deployment was that specially developed alarm schedules were in place. Even more important was the interfacing of the various departmental plans (e.g. those of the fire and medical services). The fact that the police, the authorities and others were familiar with large-scale operations also contributed to fast responses by the emergency services. Added to this, an almost complete policy staff was speedily in position and ready to advise the Mayor.

Another factor deserving mention, alongside speed, was the way external communications were established from the Co-ordination Centre. This also helped to curb the number of blatantly inaccurate reports in the media, and to limit panic in and around the area.

Turning to some of the main problems of the initial disaster response, many of our interviewees were critical of the communication links between the Co-ordination Centre and the disaster site during the first few hours after the crash. This was partly based on ignorance of the simple fact that uncertainty, confusion, contradictory reports and chaos are inevitable in this type of disaster situation. The first few minutes and even hours immediately following a disaster will always be a time of great uncertainty. On the other hand, the degree and duration of this uncertainty will depend on just how good the disaster contingency communication arrangements are. There is a technical side to this (quantity, positioning and operational potential of the equipment), but communications quality also depends on organisational factors (including compliance with procedures for logging in and radio discipline). Finally, the lack of contact between the Co-ordination Centre at City Hall and the Southeast Neighbourhood Council led to problems and unintentionally set the scene for a number of uncertainties and misunderstandings in the days that followed.

Notes

1 A.H. Barton (1969). *Communities in Disaster: A Sociological Analysis of Collective Stress Situations*, Doubleday, Garden City; J.B. Perry and M.D. Pugh (1978), *Collective Behavior: Responses to Social Stress*, 121 ff., West Publishing, St. Paul.

2 See for example: G. Kreps (1992). Foundations and principles of emergency planning and management, in D.J. Parker and J.W. Handmer (Eds), *Hazard Management and Emergency Planning: Perspectives on Britain*, 159–174, James and James, London.

3 Regional Fire Brigade Emergency Control Room, IJtunnel. Extract from the 24-hour recording, 4 October 1992.

4 R.A. van Dijk, Rampen in de Luchtvaart 'Air Disasters', syllabus symposium 'Redding en medische hulpverlening bij grootschalige verkeersongevallen' (Rescue and medical aid in major traffic accidents), Rotterdam Airport, 29 January 1993, 3.

5 The description that follows is partly based on U. Rosenthal (1988), Disaster management in the Netherlands: planning for real events, in L.K. Comfort, *Managing Disaster: Strategies and Policy Perspectives*, 274–295, Duke University Press, Durham.; U. Rosenthal and M.J. van Duin (1989), Disaster management in the Greater Rotterdam Area: Laws, Scenarios, Exercises, *Disaster Management*, Vol. 2, 2, 63–69.

6 *Brand & Brandweer*, November 1992.

7 M. Walsh (1989). *Disasters: Current Planning and Recent Practice*, Edward Arnold, London.

8 M.J. van Duin (1992). *Van Rampen Leren: Een vergelijkend onderzoek naar de lessen uit spoorwegongevallen, hotelbranden en industriële ongelukken*, (Learning from disasters), Leiden University, PhD dissertation, 19–28, 266–296.

9 See, for example: U. Rosenthal, P. 't Hart (1990). Riots without killings: policy learning in Amsterdam 1980–1985, *Contemporary Crises*, Vol. 14, No. 4, 357–376. Subsequent to 1985 there were also a number of sudden public order situations that tested police response times to the full. A notable example was the unexpected escalation of an operation to evict squatters from a building in the Tesselschadestraat in autumn 1989. This incident showed the danger that much of the operational routine and expertise built up in the period up to 1985 might have been lost. Several mistakes were made, causing the situation to escalate. As a result of this episode, a number of matters were put right and this restored the mobilisation and response capacities of the relevant organisations.

10 E.L. Quarantelli (1988). Disaster crisis management: a summary of research findings. *Journal of Management Studies*, Vol. 25, 373–387.

11 For example, compare the list of calamities taken into consideration by New York City's contingency planning: R.T. Sylves and T.J. Pavlak (1990). The Big Apple and disaster planning: how New York manages major emergencies, in R.T. Sylves and W.J. Waugh (Eds), *Cities and Disaster: North American Studies in Emergency Management*, 1990, 185–220. Springfield: Charles C. Thomas 1990, 185-220.

12 B. Raphael (1986). *When Disaster Strikes: A Handbook for the Caring Professions*, 41–68, Unwin Hyman, London. This also gives an in-depth treatment of the psychological and behaviourial effects on people of disaster reports.

13 E.L. Quarantelli (1988), op. cit. Also compare, for example: T. Shibutani (1966). *Improvised News: a Sociological Study of Rumor*, Bobbs-Merrill, Indianapolis.
14 Police Interviewee.
15 L.M. Walters, L. Wilkins and T. Walters (Eds) (1989). *Bad Tidings: Communication and Catastrophe*, Lawrence Erlbaum, Hillsdale, in particular Chapters 1, 7, 10 and 11.
16 *Haarlems Dagblad*, 6 October 1992.
17 This may help to explain why fire service interviewees systematically used the term 'major incident' rather than 'disaster'. From the angle of emergency operations the Bijlmer was largely a major accident.

CHAPTER 3

In search of victims

Introduction

This chapter looks at how the operational services and the authorities dealt with the uncertainties posed by the disaster. The major uncertainties here related to the number of casualties caused by the crash. First we discuss the initial situation as perceived by the Co-ordination Centre when the recovery operation was set into motion on Monday morning. Then we describe the course of the operation itself. Next we focus on how the police in particular went about establishing precisely who was present in the flats immediately before impact. The last paragraph summarises the most important findings from the perspective of uncertainty reduction under severe time pressure.

Coping with uncertainty: estimating the death toll

On Monday 5 October Mayor Van Thijn expressed the fear the air crash might have claimed 250 lives. Back on Sunday evening a police spokesman had mentioned 150 dead. The figure of 250 was based on the number of flat units thought to have been affected and the records of the City Population Register. On Monday the Co-ordination Centre worked on the assumption that the impact of the aircraft had destroyed 80 flat units – with 239 registered residents according to the Population Register. Although 88 residents were already known to have surfaced, possible visitors to the flats at the moment of impact also had to be taken into consideration. At that time, the figure of 250 seemed to be a realistic estimate.

The recovery operation was meant to clarify matters with regard to the number of victims. Recovery and subsequent identification of the victims would give their families and friends the certainty they so badly needed. The police were trying to discover just who was in the flats at the time. Lists were drawn up of everyone reported missing; ideally these would match the list of identified remains.[1]

Speculation

During the course of Monday more information came in that seemed to bear out the first official estimate of 250. An official of the Southeast Neighbourhood Council in the Co-ordination Centre advised that 600 people had been reported missing on the special telephone hot-line. Up to then, the police had been talking in terms of 209 persons reported missing.

According to various rumours, there were a large number of fatalities. There was said to have been a disco party in the basement or on the first floor. Large numbers of drug addicts were said to congregate in the lock-up basement storage areas. Some flats were thought to have housed between 10 and 20 illegal immigrants.

This fuelled confusion on the possible number of casualties. The idea that 250 was more likely to be on the low than the high side was reinforced by the rising number of missing persons reports. This was demonstrated in the press conference on the afternoon of Monday 5 October. Mayor Van Thijn said that estimates were rising, and that there were likely to be more than 250 bodies under the rubble. The media translated these remarks by the Mayor as 'at least 250 dead' and gave prominent place to the news. Reports in the media reinforced the idea of at least 250 fatalities. So, for example, during the first few days after the crash, it was reported that local schools feared 'between 30 and 40 victims'.[2] According to other reports, various immigrants organisations feared for many victims in their own communities; there was talk of 150 Ghanaian and some 50 Antillean dead.[3] The Antillean Premier, Mrs Liberia-Peters, had flown in to support her compatriots, who form a significant part of the population of the Bijlmermeer area. She showed up at the Co-ordination Centre with a list of 137 missing persons. Some of the other ethnic groupings represented in the Bijlmermeer also signalled large numbers of victims. All the evidence pointed to a heavy death toll.

Other signals

Even so, the figure of 250 dead was in stark contrast to the number of human remains recovered during the first few days. On the afternoon of Tuesday 6 October, 12 bodies were recovered; the fire brigade called these results 'disappointing'.[4] And so it looked as if there were plenty of remains still waiting to be recovered.

Not everyone accepted 250 as a likely figure. The city services concerned with relief and care of the survivors had a different view on the number of fatalities. By Monday, it was already clear to the Housing Department that a large number of the tenants had survived the disaster. Housing Department staff faxed a regular stream of information to the Co-ordination Centre from which it could be deduced that the death toll must indeed be considerably lower.[5] The city Social Services and the Southeast Neighbourhood Council also doubted the figure of 250.[6] From talking to the survivors, the staff of these services soon realised that far fewer flat units had been destroyed than was supposed in the Co-ordination Centre. It also became clear that many people had been able to escape from the flats that had survived the initial impact but had caught fire.

Digesting discrepancies

The Co-ordination Centre was now caught on the horns of a dilemma. While the lists of missing persons grew by the day, the ratio of bodies recovered was low in comparison. Confronted as it was with this discrepancy, the Co-ordination Centre held on to the 250 figure for a relatively long period. This policy had a definite purpose *vis-à-vis* the outside world, on the principle of 'better too high than too low'.[7]

To a certain extent, the 250 figure started to take on a dynamic of its own. A greater value was attached to evidence supporting this figure than to contradictory information suggesting a much lower body count. Information from the Housing Department that might have clarified that the death toll was far lower than 250 was not accepted at its true value in the Co-ordination Centre. Indeed, this information was only considered from the angle of allocating homes – this being the Housing Department's core task. No connection was made between these figures and estimates on the number of dead and missing. It was not until Wednesday 7 October that the Mayor and others in the Co-ordination Centre were convinced that 31 rather than 80 flats had been destroyed by the impact of the crash.

On Tuesday 6 October, two decisions were taken in order to obtain faster clarification of the actual number of dead. Firstly, the Co-ordination Centre decided to accelerate the pace of recovery. Secondly, the Amsterdam police mounted a large-scale operation to reconstruct as far as possible the actual number of missing persons. Eventually, a variety of methods came to be used to establish who was in or near the relevant flats at the time of the crash. Information from the recovery operation and the drafting of a list of missing persons were combined to give greater certainty on the number of victims. Initially the Co-ordination Centre attached greater value to the missing persons list than to the total of bodies recovered.[8] This related to a fear that some bodies had been burnt without trace. When the number of bodies failed to rise significantly, two British experts were brought to Amsterdam later during the week. They were asked about the possibility of bodies being totally turned to ashes or vaporised by the burning aviation fuel (see below).

Recovering remains

The morning of Monday 5 October marked the official start of search and recovery operations. Preparation for this had actually started during the night of Sunday to Monday. While damping-down operations were still fully underway, rescue teams of fire-fighters penetrated the flats in search of possible survivors. Danger of structural collapse prevented a search of some sections; part of the apartment buildings would need to be shored up before recovery could get underway properly.

During the night of Sunday to Monday there was also consultation between the Amsterdam city police and the National Police Disaster Identification Team (RIT). Initially, the city police had been unaware of the capabilities and expertise of this specialised national unit. Once the National Police had presented its plan of approach, the team members could start on the preliminaries for a major recovery and identification operation. At Schiphol a hangar was transformed into a temporary mortuary.

At 06.10 on Monday the fire brigade started comprehensive operations to ensure that recovery teams could work in safety, despite the difficult environment. Before recovery could start, the situation had to be approved as safe by the city Building Inspectorate. All these preliminaries took a long time. The search for remains concentrated on two locations: the area behind the blocks of flats and in the area of collapsed wreckage. Firemen carefully searched the burned out flats once again.

The aim of the recovery operations was to find human remains and to collect supplementary evidence to facilitate the identification process. Items to be noted included: location of the remains, adjacent clues and the presence of clothing. All these data were used to aid identification of the given victim. The overall identification process was supervised by a pathologist; a dentist was also involved. Identity was established by comparing information on the deceased (appearance, teeth, jewellery) with examination of, and evidence relating to, the remains.

Recovery work on the spot was performed by three crews of some 20 people, each led and supervised by an investigator from the RIT. Members of the city police serious crimes squad and the fire brigade were also attached to each crew. The recovery crews went into action whenever the fire-fighters who were clearing the rubble located human remains. After photographs had been taken and the surrounding area was searched thoroughly, the remains were removed to a large tent on the site where the city Medical Services Department performed an autopsy. Lastly, the body or body part was transferred to the mortuary at Schiphol for final identification.[9] Six bodies were recovered by Monday afternoon.

Acceleration of the recovery operation

In the Co-ordination Centre, impatience and frustration at the slow speed of recovery grew apace. On Monday the monitor had shown how the crews stopped every time remains were found, to allow the RIT experts to do their work. Based on the number recovered so far, the Co-ordination Centre calculated that it could take several weeks to recover 250 bodies.

Mayor Van Thijn and others in the Co-ordination Centre found this unacceptable. It was considered most undesirable that families and friends should be left in suspense for so long. It was also not feasible to have the police and fire brigade go on working in the area for weeks. Problems on procedures and methodology also seem to have arisen between the fire brigade and the RIT. The former felt that too much time was being lost by the team's initial, on the spot, identification work.

Thus the authorities faced a serious dilemma. The rate of recovery had to be speeded up in order to reduce uncertainty; at the same time a faster pace would increase the chance of mistakes. Indeed, lack of due care in identification might well increase the uncertainty. Although the Co-ordination Centre wanted to accelerate the recovery operation, an acceptable 'ratio of reverence' was essential. Taking this into account, the Co-ordination Centre wanted the recovery operation to proceed faster – but at the same time without loss of due care.

On the morning of Tuesday 5 October the decision was taken in the Co-ordination Centre that the recovery operation should be speeded up. It announced that the entire recovery operation had to be finished by Friday night. The Co-ordination Centre imposed a deadline on the various organisations working at the disaster site. In publicly announcing the deadline, the authorities were taking a risk. The commanders of the RIT were informed of this decision by the city police. The National Police specialists working under the command of their city colleagues loyally agreed to co-operate, but they warned that speed would affect accuracy.[10] Any mistakes would, by definition, have serious ramifications for identification.

The new methodology spurred clashes and tensions. In the opinion of the RIT, the fire brigade's pace was sometimes too fast. When this happened, one of the team's specialists would have to interfere to prevent irreparable harm to the identification process. For the RIT, the greatest hindrance in performing their task was the lack of familiarity with their work on the part of the city police and fire brigade. Again and again it had to be explained that speed must not be allowed to impinge on the exercise of due care. In the opinion of the fire brigade and the city police, however, the Co-ordination Centre's decision to accelerate recovery had a positive outcome.

Successful recovery?

The pile of wreckage began to diminish visibly, thanks to the efforts of the recovery crews and others. The RIT shifted all their activities to hangar 11 at Schiphol. The number of bodies recovered rose with the clearing away of wreckage from the disaster site.

The ground level of the destroyed flats was reached on the morning of Thursday 8 October. The fear of a discotheque area filled with bodies proved unfounded. When the recovery operations were completed on Thursday afternoon, a good 24 hours within the deadline set, far fewer than 250 bodies had been recovered. Hence, the acceleration of recovery operations gave a clearer picture of the number of fatal casualties buried under the rubble. However, the Co-ordination Centre still bore in mind that some bodies might have been totally destroyed by the intense heat. Satisfaction with the speedy recovery operation was tempered with the continuing uncertainty on the number of missing persons.

With the benefit of hindsight, many of our interviewees cited the acceleration of recovery as one of the most crucial policy decisions of the entire period. In their view, it was justified by the facts. The task was completed well ahead of schedule and the recovery operation proved not to require the many weeks initially feared.

Looking back, there are many plausible explanations as to why the recovery operation was able to be realised within the set time limit. Firstly, the decision to speed up might well have boosted the tempo on the spot. Once recovered, the remains were taken to the mortuary as soon as possible. More manpower and equipment and longer working hours all facilitated completion during the course of Thursday.[11]

A second factor accounting for the much earlier than planned completion could be that far fewer dead were found than had been expected. The original estimate had been around 250, whereas only 43 bodies were recovered, less than 20% of the expected total. In turn, this meant less time spent on recovery and related police enquiries.

Although many interviewees preferred the first explanation, in our opinion the second should also be taken into consideration. In fact, very little changed at the disaster site in the wake of the decision to speed up operations.[12] Precision came ahead of speed throughout the recovery operation. The more important factor was that many fewer 'finds' of remains meant fewer halts to the operation. If 250 had been found – as had originally been feared – recovery could not possibly have been completed on Friday evening.[13] Either there would have been a higher margin of error or the time limit would not have been met.

The experts

As early as Tuesday 6 October, the Co-ordination Centre had discussed the possibility of bodies being destroyed without trace in the inferno; this would explain the small number of bodies found. It was decided to bring in experts from abroad. According to the fire brigade, some 30% of the Lockerbie victims had never been found.[14] Professor Busuttil, the pathologist in charge after Lockerbie, was invited to Amsterdam, along with Deputy Chief Constable Orr who had supervised police operations at Lockerbie.

After a visit to the site Busuttil reported his findings in the Co-ordination Centre. He considered it likely that some 10% of the bodies would never be found.[15] The explosion would have fragmented bodies and intense heat would have totally reduced some remains to ashes. The experts also gave a second opinion on the contingency relief plan based around the announcement of the list, and various other matters.

Opinions differed on the input of the two experts. Busuttil's report contained the following statement: 'It is my considered opinion that it would be highly unlikely that as many as 100 bodies or more could have perished completely in this particular disaster by incineration to ashes or otherwise.'[16] His analysis did contribute to a more considered estimate of the death toll, albeit very much on the side of prudence. Further, the Scottish expert opinions supported the view that great care and sensitivity would have to be exercised *vis-à-vis* the survivors and families. This input was greatly valued by Mayor Van Thijn.

Some operational officials were less pleased – the impression was being created that the Netherlands did not possess the relevant expertise. Moreover, the recommendations made by the two British experts meant a lot of extra work. Busuttil's advice that, for the sake of certainty, a large area around the disaster site should be excavated, was greeted with little enthusiasm, particularly in fire brigade circles. Moreover, as it later transpired, it was most unlikely that any bodies had been burned to ashes or vaporised.[17] Also, the arrival of the experts proved a great media attraction. This distracted some critical attention from other developments in Amsterdam that were proceeding less favourably at the time, like the relief work in the Bijlmer Sports Centre.

Identifying the missing

The second strategy used to reduce uncertainty and establish the precise number of victims consisted of reconstructing the situation in the flats at the moment of the crash, for example, to establish who could reasonably be expected to have been present in or around the apartments at the time. This proved a difficult task for the Amsterdam police, working together with organisations including the City Population Register and the Southeast Neighbourhood Council.

By the afternoon of Monday 5 October, the number of missing persons reports had passed the 600 mark. On the morning of Tuesday 6 October, the senior policy-makers in the Co-ordination Centre decided to issue an appeal for re-registration of persons reported missing. This appeal was duly broadcast on local TV, in various languages, which produced another flood of reactions. During the course of Monday, the police formed a special disaster enquiries team to check on how many people were really missing. By Tuesday evening the number reported missing had risen to 1,100; it would later rise again to 1,588.

Although everyone was convinced that it was simply impossible for so many people to have lived in the wrecked flats, it was still unclear how many of those reported missing were in the vicinity at the time of the crash. The team of police investigators was instructed to get a clear picture of the number missing, as soon as possible.

Explaining the large number of missing persons

There are various explanations for the extreme size of the list of missing persons. Some individuals used the disaster for personal benefit. One creditor even saw it as a way of obtaining the address of someone who owed her money. Even so, without devoting much valuable police time to checking this type of report, there was no way of safely establishing which were and were not false alarms. Added to this, those reported missing came from dozens of different nationalities; this compounded the problem of drafting the list. There were multiple spellings of the same name to cope with, so that one particular person might be listed twice or more. Many names were spelled phonetically, and many foreigners were known by a nickname or alias. Particularly in the case of Africans, it transpired that different generations used different names for the same person. The presence of illegal immigrants among the missing did not help matters.

Furthermore, it appeared that tiresome procedural mistakes had been made in compiling the lists, particularly during the first few days. When the police investigators got to work on Tuesday, the first thing they had to do was to impose some order on the procedure followed to date. The padding of the list had been partly due to the methods used to register missing persons by Southeast Neighbourhood Council staff in the Bijlmer Sports Centre during the first few days. Reasonably enough, they had not been trained in this specific task, and so basic data had been missed or entered incorrectly. As a further example of what had gone wrong, there were no meaningful, standard forms on which to note all the relevant information. Although missing persons reports were accepted at several locations, for the first few hours there was no collation or comparison of incoming data. 'Found' reports were not always properly processed; telephone numbers were noted incorrectly. The lists contained many duplications; one family of four appeared no less than 17 times on the lists. The officials often failed to note the name of the person making the report, so that there was no way of verifying details, and no way of telling the caller when the missing person was found.

Creating order out of chaos

On Tuesday 6 October the Southeast Neighbourhood Council personnel who had been handling telephone calls from the public about missing persons until that time handed this task over to police investigators. Using an improvised standard form, police imposed uniformity on the registration of missing persons. All the same, the police team had to deal with an enormous backlog and their first task was to rationalise the list: duplications had to be removed and differentiation was made between missing persons who lived in the actual disaster area, and those from outside it. The Director of the City Population Register seconded personnel to help the police with this work.

Imposing order and system on the work of the first 36 hours was not the only task facing the police enquiry team; additional work landed on their desks. All the time they were busy screening the existing list of missing persons new reports continued to flood in. In view of the numbers already reported missing, the Co-ordination Centre decided to halt the appeal via the local TV station.

A considerable extra workload was the result of interventions from the government of the Netherlands Antilles. On Wednesday 7 October the Co-ordination Centre gave the police enquiry team a list of 137 missing persons of Antillean origin that it had received from the Netherlands Antilles Premier, Mrs Liberia-Peters. This list also needed vetting. The list of missing Antilleans eventually rose as high as 178. An Amsterdam police liaison officer was sent to the Netherlands Antilles. The list was soon reduced to zero.

Increased time pressure

By Wednesday night the list of persons reported missing had been checked and processed into a more reliable document. The original list, which at one time had risen as high as 1,600, had been reduced to 300 names, all of whom could realistically be considered as missing.

On Wednesday evening it was decided in the Co-ordination Centre that a list of identified fatalities and a definitive list of missing persons would be released on Friday 9 October; it was considered unacceptable to keep grieving families in suspense any longer. This decision intensified pressure on the police.

On Thursday morning police investigators, numbering more than 50 at that stage, set to work on the basis of the list of 300 likely missing persons. For the police, finding missing persons on the basis of the correct information was a matter of routine, 'good old-fashioned' detective work.[18] Priority went to people said to have been living in the Bijlmer. The investigators made house-to-house enquiries to see if the missing were still alive.

With these house-to-house enquiries, the pace started to quicken. Although consulted, the City Population Register proved to contain insufficient reliable information to establish who had lived in the flats. Detective work was needed to reconstruct the situation in the flats at the moment of impact. Who lived in which flat? Who was present? Were there visitors? Names had to be found for every single flat. Checking on the inhabitants would establish who had survived the disaster. As information from the population register alone had proven inadequate, the police also had to use lists and records from the telephone company, the Social Services and the Nieuw Amsterdam Housing Co-operative, as well as less obvious sources like mail-order companies. At one point the number of lists consulted by the police investigators actually rose to 34.

During the first week much time had to be devoted to pruning then checking through the lists of persons reported missing. Considerably less time went into the 'reconstruction method', partially due to the tight schedule imposed on the enquiry team by the Co-ordination Centre. The authorities in the Co-ordination Centre wanted to see a 'definitive' list of identified fatalities and missing persons by Friday afternoon.

Release of the lists: X-Hour

The time chosen to release the names of the dead and the missing was set some what arbitrarily. Preliminaries took up Thursday and most of Friday. Mayor Van Thijn was anxious that nothing should go wrong and that every care and consideration should be exercised when releasing the lists. The time and place for this were to be Friday 9 October at 17.00 hours, in the Bijlmer Sports Centre.

This event dominated the whole of Friday. The operation proved to require an enormous organisational effort. And the need on the Friday morning to make decisions on dozens of matters – while none of these was overly complicated – when taken as a whole put heavy pressure on a few people. These decisions included:

- The time to be set for X-Hour (release of the lists);
- Arrangement for emotional support of family and friends;
- Whether or not to divide up the names by ethnic group;
- How and when to inform the press;
- How to arrange for production of 3,000 lists for distribution in the sports hall within a few hours;
- How to get everyone to the Sports Centre and then back to the overnight accommodation;
- Who would man the permanent telephone lines after the lists were released.

The press conference started at 18.22, a good hour later than planned. Shortly before, the main national TV news had broadcast inaccurate information about the number of dead. Notwithstanding, matters proceeded calmly in the sports hall (see Chapter 4).

However, a more severe problem was that the lists had really been released prematurely. Due to the very tight schedule imposed, several mistakes had crept into the lists. The police enquiry team had needed all available manpower to meet the Co-ordination Centre's wish that unverified lists of missing persons should be refined down to a single list before Friday afternoon. The Co-ordination Centre was insufficiently aware of the constraints facing the enquiry team. The team had to work to the highest level of painstaking accuracy: no name which had already been identified could be allowed to slip back into the missing list.

There was simply not enough time for the police to complete a definite list of missing persons before Friday afternoon; and so, on Friday morning it was decided that the list would be split into two, differentiating those missing persons who were likely to have been in the disaster area, and those who were not. Added to the list of already identified bodies there were now three lists.

The police enquiry team was aware that the lists were only an interim report. Lack of time had prevented them from checking on the actual residents of the relevant flats. So, for example, it was known that a number of Ghanaian illegal immigrants had been living in the wrecked flats; hence it was quite possible that some victims might never have been reported missing. Further investigations would be needed to clarify this. New, altered lists were produced within hours of the initial lists being released. And, day by day, successive lists became more reliable.

Conclusions

The task completed in Amsterdam between Sunday 4 and Saturday 10 October was a difficult one, requiring a high degree of organisational and improvisational ability on the part of the people and institutions involved. Moreover, they were closely monitored by the media and in turn the entire Dutch population.

For a large part of that first week, the Co-ordination Centre and certain organisations continued to attach great value to the original estimates of the death toll (i.e. 250). Information seeming to support this figure was accorded far greater value than apparently contradictory news. Also in this period considerable confidence was invested in the high total of missing persons, while the small number of bodies found mainly prompted uncertainty.

In addition, many of the problems with the lists of missing persons had their roots in the organisational difficulties of the very first few hours and days. The organisations involved at that time were unprepared and had had to act as best they could. This created a lot of extra work later on.

The Co-ordination Centre imposed two deadlines on the organisations involved; this had both positive and negative ramifications. On the one hand it spurred them to fix a delivery time when the outside world could be clearly informed on the situation. On the other hand it increased the pressure on already stretched resources and personnel. And, while designed to reduce the suspense, these deadlines simultaneously created the danger of one variety of uncertainty being replaced by another. Any mistakes made during the recovery could be irreversible. The speedy publication of the lists increased the chance that people who had already been found would still be noted as missing.

Notes

1 With hindsight this is quite rational. At the time the officials of the Southeast Neighbourhood Council and the police started to check the names of the missing persons, no thought had been given to congruence of this list and the list of identified bodies.
2 'Fear and hope in Bijlmer schools', *Het Parool*, 6 October 1992.
3 'Visit from Antilles' and 'Fear of high Ghanaian death toll', *Het NRC Handelsblad*, 6 October 1992.
4 Co-ordination Centre log book, 34.
5 Housing Department interviewee.
6 Source: interviewees and reports from these organisations. However, these reports were drafted well after the events.
7 Source: several interviewees from various services in answer to the question why this figure of 250 was maintained for so long.
8 'List of missing will be definitive: not the identification', Co-ordination Centre log book, 27.
9 In the first instance it was possible to establish the identity of 41 people. DNA tests in the United States later established the identities of the last two fatal casualties.
10 National Police Disaster Identification Team (RIT) interviewee.
11 A view taken by fire brigade interviewees.
12 RIT interviewee.
13 RIT interviewee.
14 Co-ordination Centre log book.
15 Co-ordination Centre log book, 72; Prof. Busuttil's report, 3.
16 Prof. Busuttil's report.
17 See Chapter 1, note 12.
18 Labelled as such by Deputy Commissioner Bernard Welten at the Friday press conference.

4 October 1992: blazing ground fires in the Bijlmer shortly after the crash. (Photo: ANP)

The press cluster around victims in front of the Bijlmer Sports Centre. (Photo: ANP)

Recovery and identification in full swing: Disaster Identification Team (RIT) police and fire brigade. (Photo ANP)

Fire-fighters at work amidst the wreckage of the jumbo jet and the collapsed apartments. (Photo: ANP)

Fire Chief Ernst addressing Amsterdam officials at a briefing session in the Co-ordination Centre. Standing behind him: Sarucco, Head of Public Order and Security. Seated in the front row, left to right: Deputy Director Hoff, General Administration and Legal Affairs Department; Mayor Van Thijn; Van Ostreen, Head of Information. Seated behind the Mayor: City Council Secretary Kooiker. (Photo: City of Amsterdam)

The aircraft crashed into 'Groeneveen' and 'Kruitberg', two linked apartment blocks (Photo: Korps Landelijke Politiediensten)

A welcome cup of coffee. Support for the police, fire-fighters and other regular services from the Salvation Army and other voluntary organisations. (Photo: ANP)

Briefing in the Co-ordination Centre, Police Superintendent Welten gives the latest update on the missing persons enquiry. The TV monitor provides the Centre with direct on-going coverage of the recovery operation at the disaster site. (Photo: City of Amsterdam)

Mayor Van Thijn signs the condolence register. (Photo: ANP)

Consultation underway in the Co-ordination Centre. Fire Chief Ernst and Chief of Police Nordholt in the middle of the front row. At the end, right: Mayor Van Thijn. (Photo: ANP)

The Bijlmer Sports Centre: distribution of the lists of identified victims and missing persons. (Photo: ANP)

Expression of mourning and solidarity at the disaster site. (Photo: ANP)

Official memorial service in the RAI. Members of the Dutch Royal Family, Government and Amsterdam City Council, with foreign dignitaries. (Photo: ANP)

The bitter dividing line of a disaster. (Photo: ANP)

Illegal aliens queuing outside the Population Register.
(Photo: ANP)

A gap in the Bijlmer: demolition work at the disaster site.
(Photo: Korps Landelijke Politiediensten)

CHAPTER 4

In aid of survivors: psycho-social care

Introduction

At a single stroke, a disaster like the one in Amsterdam can convert large numbers of people into victims requiring help. Such destruction, coming without warning, robs people of their sense of security and confidence in the routine and predictability of life. This applies to everyone who experiences the impact, whether they lose someone dear to them, or their home or belongings, whether they are injured or lucky enough to escape with shock. In this state of upheaval many of those affected will need help. We distinguish between two categories of help: the material and the psycho-social. This chapter describes how the relevant agencies went about providing help and support to those affected by the Bijlmer disaster. The emphasis is on psycho-social help, which we take to mean all intangible forms of help, including psychological care and social support. In particular we will be looking at the organisation of relief and after-care, rather than the psychological impact of long-term care on the victims.[1] Material relief, i.e. tangible and financial assistance, including cash grants and housing, is described in Chapter 6.

Victim profiles

Relatively few of the victims were native-born Dutch. Most were immigrants, hailing from every continent and region. Their places of origin included the Netherlands Antilles, Aruba, Surinam, Ghana and other African countries, South America, Turkey, India, Pakistan, and several parts of the Arab world. The specific composition of the group of victims had important ramifications for the relief and support required. The people involved were relatively socially disadvantaged and – as one relief worker remarked – some of them '.... were totally lost' in the bureaucracy of Dutch social benefits and support.[2] Moreover, some of them were already traumatised prior to this air crash by events in their own countries, like political persecution, or – as in the case of some people from Surinam

– by involvement in the crash of a Surinam Airlines jet at Zanderij Airport (in Surinam) in 1989. The fact that a number of the victims were illegal immigrants in the Netherlands did not make matters for them or the relief workers much easier.

Initially, victims lacking regular immigration status were reluctant to go to official relief centres, for fear of being registered. Many sought and found a roof over their head in their own communities or in a local church centre. At a very early stage the Co-ordination Centre had stated that all victims should receive equal treatment, no matter what their legal status.[3] The authorities made great efforts to ensure that the 'illegals' also received the help to which they were entitled. At the same time it has to be borne in mind that for the authorities this was not exclusively a matter of providing help. They had another interest: obtaining as comprehensive as possible a picture of the number of victims of the disaster. Information from and on the illegals was crucial for this. Acquiescing to the fact of 'anonymous dead' was considered morally unacceptable by members of the Co-ordination Centre.[4]

The relief centres

In the first few hours after the disaster, the city authorities set up relief and reception centres at various locations. Numerous schools, churches and neighbourhood centres also spontaneously opened their doors (see the map of relief centres on p. 61). Many of the newly homeless found accommodation with families or friends in the immediate vicinity.

Thanks to this wide availability of reception and relief facilities, there were no difficulties in providing accommodation and help for the many people affected. However, the geographical dispersion of relief centres did bring its share of problems. Firstly, the distance between them complicated co-ordination and exchange of information between the operational services. This problem was exacerbated by poor communications – which also hampered contacts with the Co-ordination Centre. Moreover, those involved in providing immediate care and transport for the victims in the first few hours after the crash were confused as to where they should be taken. Worried friends and family were also milling around the disaster site, unsure where to go. Another problem was the spread and consequent fragmentation of care and aid activities. This hampered registration of the victims, which in turn made it difficult adequately to inform family members on the fate of the people living in the flats.[5]

During the evening of the crash, the police, using dispatch riders, obtained sufficient information on the people present at the various reception locations, and then took the decision to centralise relief operations.[6] The Bijlmer Sports Centre was the most suitable place, in terms of size and proximity to the disaster site. Evacuees from the smaller reception and relief centres were then transferred to the Bijlmer Sports Centre. Some 200 evacuees and their family and friends were looked after on the first evening.

The Bijlmer Sports Centre

Acting on his own initiative, the manager of the Bijlmer Sports Centre had opened his doors for emergency relief of the victims of the disaster. A few hours later, the necessary infrastructural provisions like telephone lines and computer equipment were arranged and installed under police supervision. The Sports Centre was now easily reached by telephone and, at a very early stage, was adequately equipped to cope with the relief of a large number of victims. All the same, according to some people, initially the situation in the Bijlmer Sports Centre was somewhat confused. It took a full day before sufficient structure had been imposed to show clearly where people should go with specific questions or problems.[7]

On the first night the help available outstripped demand: there were 70 relief workers and 90 victims. One explanation for this is that large numbers of both voluntary and full-time relief workers had streamed to the reception and relief centres, while the people affected had sought and found accommodation and help elsewhere on the first night. This situation, with a high proportion of people making their own arrangements so that there is relatively little call on official facilities, is quite common in disasters.[8]

The provisions were further expanded in the days after the crash. Desks for the representatives of the various nationalities involved and the relief organisations were set up in the main sports hall. This gave a clearer picture of the composition of the group of victims and the presence of the various services and organisations (see plan of the Bijlmer Sports Centre on p. 63). The various relief agencies also did their best to meet the specific preferences and customs of the victims – for instance by delivering appropriate food.

After a week some of the organisations started to reduce the level of provisions and services. During the second week, several agencies like the city Social Services and Housing Departments were only present in the Sports Centre for a few hours a day. After the inter-faith memorial service on Sunday 18 October, relief activities closed down in the Bijlmer Sports Centre. Anyone needing further help could use two new facilities: an information and advice centre and a walk-in centre. The provisions were reduced gradually so that no one involved would get the feeling of having been abandoned.

Mass convergence

Immediately after it had become operational, the press arrived at the Bijlmer

Sports Centre *en masse*. As the police kept them outside, the journalists turned their attention to victims who were still outside the relief centre. VIPs, dignitaries and senior officials formed another category of visitors. The value of their presence for the victims should not be underestimated, but preparing for and coping with these visits meant an additional burden on the police and other relief workers.

Overnight accommodation

On the first night, the city authorities had arranged sleeping accommodation at several locations. The city Social Services co-ordinated the accommodation. Buses shuttled people back and forth between the day care and overnight centres. Much improvisation was required in the first week. Initially there was a large amount of overnight accommodation available, including a naval barracks, hotels, youth hostels, churches, nursing homes and refuge centres, which had been offered quite spontaneously. Most were not needed as the city's own capacity proved sufficient. Two city facilities were used for overnight accommodation, from Monday evening onwards: the naval barracks and the merchant marine academy (see the map of relief centres on p. 61) – with a joint total of 250 beds. As with the Bijlmer Sports Centre, as far as possible the dignitaries and press were steered away from these overnight locations.

Manageability of the various relief centres became a problem during the course of the week. It seemed difficult to get a good overview of the victims needing relief and accommodation, as the composition of the group and its numbers varied from night to night.[9] According to relief workers, the initial atmosphere in the centres was one of resignation, but during the course of the first week the mood among certain ethnic groupings became much more demanding. At the barracks in particular there was some tension between victims and naval personnel. Given the differing cultures and daily routines of the temporary visitors and their hosts, this was hardly surprising.

Overnight accommodation at the naval barracks was halted at the end of the first week, and the city had to seek alternative locations. Eventually the Novotel and Park Lake Hotel were selected; the latter had already been reserved by the Antillean authorities for members of their community (see Chapter 5). With benefit of hindsight it is unfortunate that those needing relief were spread among so many different overnight locations, which were, moreover, far away from the Bijlmer Sports Centre. It hampered a meaningful overview of the group affected, making it easier for people not properly entitled to benefit from the overnight facilities. To prevent this abuse, those receiving relief had to show an ID every time they went into one of the centres. This meant several ID checks every day, which proved burdensome for all parties involved.[10]

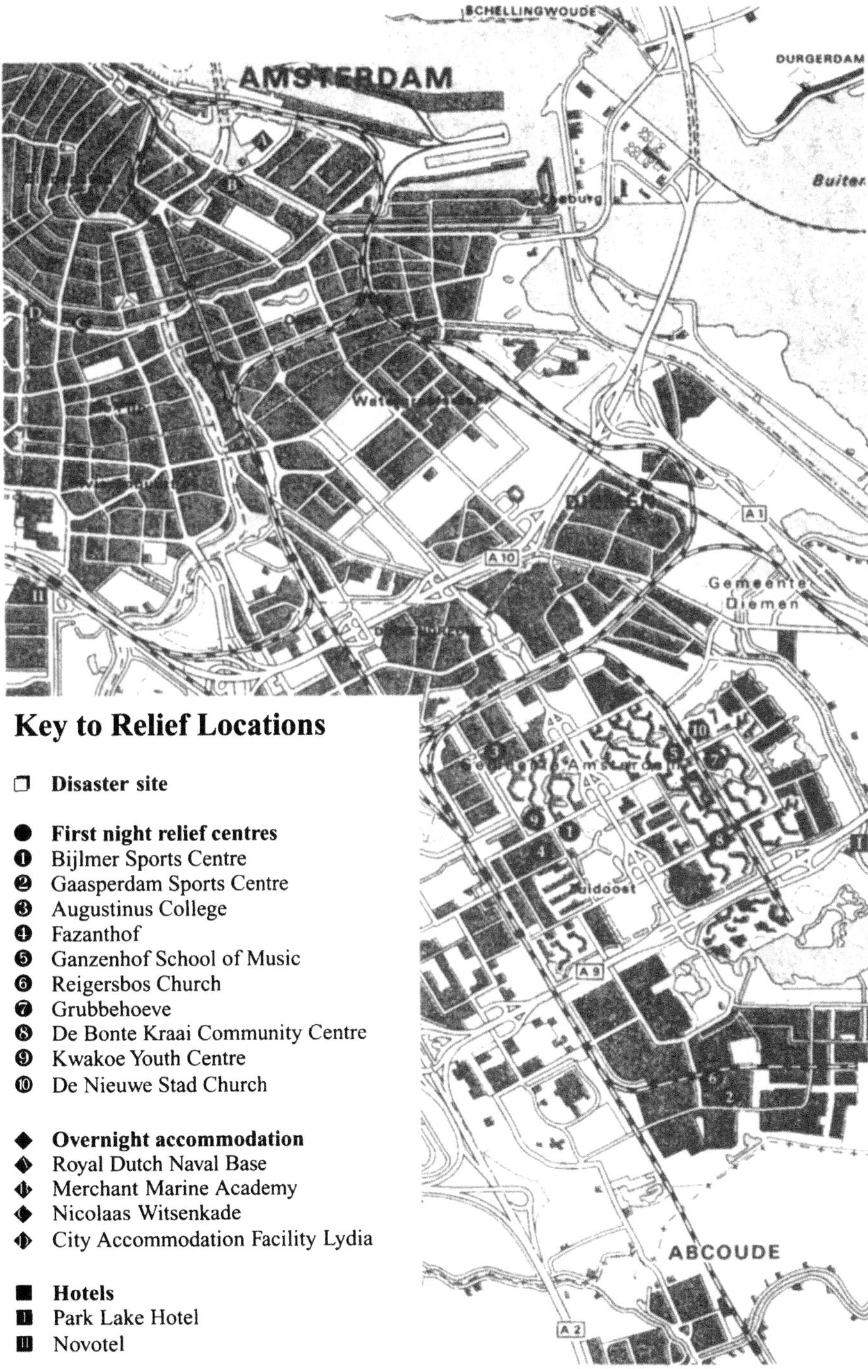

Key to Relief Locations

❒ **Disaster site**

● **First night relief centres**
❶ Bijlmer Sports Centre
❷ Gaasperdam Sports Centre
❸ Augustinus College
❹ Fazanthof
❺ Ganzenhof School of Music
❻ Reigersbos Church
❼ Grubbehoeve
❽ De Bonte Kraai Community Centre
❾ Kwakoe Youth Centre
❿ De Nieuwe Stad Church

◆ **Overnight accommodation**
◆ Royal Dutch Naval Base
◆ Merchant Marine Academy
◆ Nicolaas Witsenkade
◆ City Accommodation Facility Lydia

■ **Hotels**
■ Park Lake Hotel
■ Novotel

Admission and registration

One important condition for proper adaptation to traumatic events is that families and communities should be kept together as much as possible, and that evacuees are carefully registered.[11] The former was taken into account sufficiently during the evacuation, but registration at the reception and relief centres was faulty. In turn, this hampered enquiries by families and friends.[12]

The police ensured that the press were prevented from entering the Bijlmer Sports Centre and harassing victims. While this made good sense, it had the bizarre effect of making it relatively difficult for white victims to gain admission to the reception and relief centre. Native Dutch people who could not prove that they were relief workers were sometimes refused entry on the first evening because it was assumed that they were journalists.[13]

Co-ordination in the Bijlmer Sports Centre

The rich variety of people needing help in the Sports Centre was matched by the large number of relief organisations present. Provision of the help and support needed required the co-ordinated deployment of several, very different municipal and private organisations. These ranged from the police to priests, from regional psychiatric care agencies to the Salvation Army, and from the city Housing Department to more or less formal ethnic interest groupings. Given the diversity of experts, cultures and interests represented by these organisations, co-ordinating their activities was no easy task.

Consultative structures

Co-ordination was arranged informally on the first evening and night, with the police as the central point. But spontaneous co-ordinating activities by people not usually encountered in disaster planning, like the manager of the Bijlmer Sports Centre and several resourceful volunteers, also proved crucial. For the first few days it remained somewhat uncertain precisely which body bore political and administrative responsibility for matters in the Bijlmer Sports Centre. After a while, the Mayor indicated that the Southeast Neighbourhood Council would bear formal responsibility for matters in the Sports Centre.[14] Up to that point the Council had not had a high profile in the Sports Centre, and when its executive officials tried to assume authority they found that the police were already very much in control of things. For example, a consultative co-ordination structure had already been established under police chairmanship, and with the participation of organisations on the spot. This consultation took place every couple of hours, and later twice a day.[15]

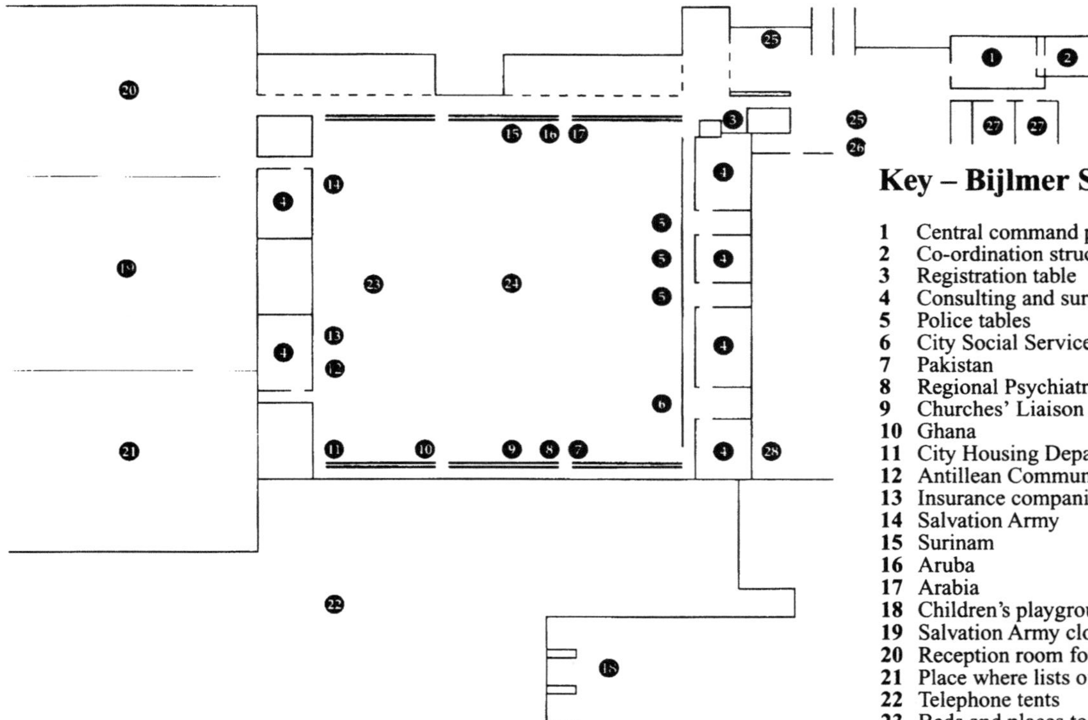

On the first floor (not displayed on this map) was a creche, prayer room, a room for the police telephone team and a meeting room for official delegates.

Key – Bijlmer Sports Centre

1 Central command post
2 Co-ordination structure first night
3 Registration table
4 Consulting and surgery room
5 Police tables
6 City Social Services. City Population Register
7 Pakistan
8 Regional Psychiatric Aid Agency
9 Churches' Liaison Council
10 Ghana
11 City Housing Department
12 Antillean Community
13 Insurance companies
14 Salvation Army
15 Surinam
16 Aruba
17 Arabia
18 Children's playground
19 Salvation Army clothing store
20 Reception room for official representatives
21 Place where lists of dead and missing released
22 Telephone tents
23 Beds and places to sit
24 Television tower
25 Information counter
26 Telephone counter
27 Consulting room
28 Meeting rooms for informal groups

Alongside this consultative structure, a similar process was set up during the course of the first week for representatives of the various ethnic groupings – or 'country teams' as they were called.[16] The idea was that these country teams would be used to channel requests for help and break through language barriers. Requests for help were referred from the country teams to the consultative co-ordination level, so establishing a match between supply and demand. To keep matters manageable there was one representative per country or language area, despite the fact that a given country might have many different ethnic groupings. These representatives were nominated by the chairpersons, who '.... tried to fish out the people with a positive attitude'.[17]

It did prove difficult for the Neighbourhood Council to make good the status shortfall as regards co-ordination and management. All sides continued to accept the police in the chair of the co-ordinating consultative structure. Apart from this, the Council was also mainly concerned with implementation and less with policy making, the trouble surrounding the pass arrangements being a good example of this.

Criteria for entitlement to help: the passes

On the first night the total number accommodated was around 200. This number rose sharply within a few days and suspicions grew that pseudo-victims – 'free riders' as they were called – were abusing the facilities offered. The Social Services and the Salvation Army also recognised a number of their clients from the city centre. This prompted the Co-ordination Centre's decision to introduce a system of passes: relief workers and victims alike would have to show ID to get into the Bijlmer Sports Centre. The Southeast Neighbourhood Council was instructed to implement the decision. This proved difficult. To start with, it was hard to formulate definite criteria for entitlement to help. The criteria used by the various services for entitlement to day care in the Bijlmer Sports Centre were highly subjective, varying from 'we know them, they know us' to 'their story sounds serious'.[18] The term 'relief worker' was also open to multiple interpretation. Added to this, when it came to implementing the pass system, each service involved had a different interpretation of the agreements made in the Co-ordination Centre. Also, passes issued at the overnight accommodation locations were already in circulation.[19]

It was agreed in the Co-ordination Centre that the Southeast Neighbourhood Council would draft criteria for the issue of passes, in consultation with the police and other services. The parties were unable to reach a consensus on this. There was also confusion on where and when passes would be issued. The Neighbourhood Council and the city Social Services also had differing views on which categories of victims should get a pass: should this be everyone present in the Bijlmer Sports Centre, or just family and friends of those directly affected? All in all, both helpers and the helped were unsure who needed a pass and how this could be obtained.[20]

The Council eventually issued passes to a large number of people. The Council officials did not feel that they could apply any meaningful criteria to who was, or was not, a victim or relief worker. Moreover, they faced long queues of often very angry victims, relief workers and others at the Council offices. The Council simply did not have the infrastructure or manpower to process them fast enough. The riot squad had to be called in to maintain order.

Although police were controlling access to the Bijlmer Sports Centre, they soon realised that almost everyone had a pass.[21] The recipients' details had been registered prior to the passes being issued. This had made some illegals reluctant to report to the Sports Centre, and so, after a few days, the Council, in consultation with the police, decided to issue day-passes in the form of adhesive labels; one didn't need to register to get these.

Some form of control on access was needed to prevent abuse of the facilities and benefits. At the same time it seemed likely that there would be considerable opposition to any attempt by the authorities to exclude certain people from the relief efforts. In an underprivileged area like the Bijlmer, disaster can easily spur a perception of second-class treatment. And the post-disaster feeling of sharing a common fate can draw the victims closer together. Unrest and aggression flourish in this climate, and a sensitive approach is called for. This may place heavy demands on the versatility of relief workers.

As envisaged, the pass system had a deterrent effect, but it should have been more carefully prepared. The Co-ordination Centre was probably unaware of the practical difficulties that would arise in implementing and enforcing it. More generally, in the first week the Co-ordination Centre had little idea of exactly what was happening in the Bijlmer Sports Centre. Most of the authorities' attention was focused on developments relating to the disaster site and the lists of missing persons.[22] The Co-ordination Centre was also insufficiently aware that the administrative status quo deprived the Southeast Neighbourhood Council of the authority to make city services toe the line.

Co-operation between relief organisations

The idea was to offer 'integral care and support' in the day centre, i.e. psychological and other medical treatment, and social and tangible support. During the first few days the services and organisations in the Bijlmer Sports Centre made an enormous joint effort. It was seen as a common goal, and people did whatever was needed, with no regard for inter-service rivalry or office hours. It is a well-established fact that the longer the relief effort continues, the more this togetherness comes under pressure.[23] After a while, there is confusion on roles and tasks; fatigue and emotion start to take their toll. This was manifest in several ways during the Bijlmer disaster.

To start with, the various players differed on the type of help needed. Did the people affected need immediate information on insurance and new housing, or was the prime need for psychological support? Did the Bijlmer Sports Centre have a purely symbolic or a purely instrumental function? Bodies like the South-east Neighbourhood Council, ethnic organisations and the Co-ordination Centre saw the Bijlmer Sports Centre partly as a meeting point for people in a similar predicament. The Co-ordination Centre in particular emphasised the importance of integral care and support, with material and psychological aid being administered under the same roof – to all ethnic groupings. In contrast, the city Social Services Department defined the day care facility as an advice and information centre, and took the view that islands of care had been formed in the Bijlmer Sports Centre, inhabited by organisations which had only their own target group's interests in mind. These differing viewpoints were demonstrated in the discussion about closing down the Bijlmer Sports Centre.[24] Views also differed on approaches to the multi-ethnic factor. Should those affected be treated as a single group or should the approach be customised? And in how far could 'white' relief workers help victims who were not Europeans?

At first, the desirability of 'white' or 'black' relief workers did not appear to be an issue for all ethnic groups. But the Antillean community in particular wanted to have nothing to do with white helpers at first. Mistrust of the authorities ran very deep.[25] This would ebb later on, but there were several communication breakdowns between the authorities and the victims. People were upset by all the forms to be completed, and the requirement to show their papers when their homes had evidently been burned out. Western-oriented psychological support provided by the city took a more 'listening', less concrete and less prescriptive approach than in some other cultures. These differences in approach were, however, not unbridgeable. There were plenty of relief workers with non-Dutch backgrounds, from both the city services and other agencies and groups. Moreover, the country teams' co-ordinators acted as a link between the various cultural groupings.

Another issue was how to cope with tensions in the various relief centres. Is anger a normal post-traumatic phenomenon? Would police intervention diminish or fuel the chance of escalation? Both the Bijlmer Sports Centre and the overnight centres saw incidents whereby groups of victims felt they had been treated improperly and showed this verbally or with threats of or actual physical violence. When this happened, the relief workers tended to call on police assistance. The police, however, adopted a cautious stance for fear of heavy-handed action causing an escalation in such an emotionally charged atmosphere. Often threats proved manageable and matters could be solved by talking them out.

Special forms of care

Alongside the care provided in the reception and relief centres and hospitals in the period shortly after the disaster, the city authorities also implemented a number of measures specifically tailored to the situation.

Release of the lists

The city authorities set themselves the task of reducing the uncertainty on the number of dead as soon as possible, and of being able to inform survivors on the fate of their families and friends. Lists of dead and missing compiled by the police were released on Friday 9 October (see Chapter 3).

Expecting large crowds and strongly emotional reactions when the lists were released, the Co-ordination Centre decided that matters should be arranged with the greatest care. Aided by outside psychiatric experts, a scenario was drawn up, featuring the time and place of the release, who should make the announcement, and in which language, and whether or not the news should be issued by nationality or collectively, whether verbally or in writing.

At 'X Hour' the Sports Centre was divided into sections so that each ethnic group could receive and assimilate the news among its own people. The co-ordinators distributed the lists and information, in the relevant language, to the various groups. The entire operation was also 'covered with care workers'.[26] A large tent had been set up, with telephones for the victims and people involved to call relatives.

In the event, when the lists were released, the expected flood of emotion failed to materialise. In the days leading up to the announcement, people had been well prepared for what was coming. For many, the certainty provided by the lists brought relief. Some of those directly involved felt that the stringently managed announcement procedures and the surrounding fuss contributed to an unnecessarily loaded atmosphere.[27]

The memorial services

The official service of commemoration was held in the RAI exhibition centre, in Amsterdam, exactly one week after the disaster, preceded by a silent procession. As far as possible, preparations for the ceremony took into consideration the wishes and customs of the various cultures. Fifteen organisations came forward on behalf of the ethnic groups involved; together with the city and Neighbourhood Council authorities, they made up the Bijlmer Air Disaster Memorial Service Committee. The ceremony organised by the committee was necessarily the largest common denominator of the wishes of the various communities. Most of the organisations and groupings involved also arranged their own services of thanksgiving and remembrance, alongside the city's ceremony in the RAI.

Sunday 18 October, the day of the big ceremony in the RAI, was also the Bijlmer Sports Centre's last day as a care and support facility. A religious service was held to mark the closing. A temporary memorial was established at the site of the crash, later to be replaced by a permanent monument.

Flights banned

Out of consideration for the residents around the disaster area the Co-ordination Centre reached agreement with the Ministry of Transport and Public Works that there would be a ban on overflights up to and including the day of the memorial service, Sunday 18 October. However, on several occasions, weather conditions meant that this route did have to be used – not surprisingly to the dismay and anger of local residents. The flight path over the Bijlmer was gradually reopened following the religious service on 18 October. A number of local residents attempted to obtain a court order banning these flights, but failed.

After-care

Shortly after the crash, the Co-ordination Centre instructed the head of the psychiatric section of the city Medical Services Department to draft a longer-term after-care plan. This had not been provided for in the city's disaster contingency plan. The after-care plan, incorporating psychiatric expertise, was ready within a week.

The aim of after-care was to avoid psychological damage by providing integrated care, i.e. linking material and psycho-social help. In addition, it was decided that after-care should be provided as much as possible by the victims' and relief workers' existing social networks and organisations. It was thought likely that most of those affected would get adequate help in the form of social support from within their own milieu, given that the strain experienced is a normal reaction to a traumatic event. The after-care plan comprised a broad range of care and information, right through to psychological diagnosis and therapy.

Integral help and support in one's own milieu was a strong starting point. This was backed up by experience gained after the Surinam Airlines disaster. But there were risks attached. Due to the many different organisations and bodies involved, this approach demanded intensive co-ordination and progress monitoring. The regional psychiatric care agencies were relatively autonomous professional organisations, not used to operating under central direction.[28] Moreover, the ethnic organisations and the country teams, which were meant to provide care within their own circles – according to the after-care plan – did not possess the full capability for this.[29] Matters were further complicated by the fact that the victims had moved to new addresses spread over a far greater area than Amsterdam Southeast.

There was also the problem that some of those affected (for example from countries like Nigeria and Togo) could not rely on strong social networks.[30] The after-care plan stated that these people should be served by professional care sources. The idea was to involve ethnic care workers. Looking back at the episode later, some people felt that to date this has not happened often enough.[31]

In order to cope with this type of problem the plan provided for a care team to make house calls and check on the need for help on the part of those directly affected. By spring 1993, 64% of the directly affected households had been reached. This met the need for a 'reaching out' type of care – at least as far as those directly involved were concerned – whereby care providers actively seek out the victims.[32] Following the house call the after-care team referred people to the regular care providers, and acted as intermediaries if need be.[33]

Conclusions

Many municipal service and other groups and organisations made a tremendous effort to provide psycho-social support to the victims, their families, and the community as a whole. Attempts were made to customise care to the specific requirements of a multi-cultural situation. Even so, there was room for improvement.

Providing relief, support and care for people hit by disasters makes organisations and services work together far more intensively than in a normal situation. Moreover, some organisations and services may also have to cope with new and unfamiliar roles and tasks; at the Bijlmer disaster not everyone was equally well prepared for this. In the prevailing institutional configuration at the time, the Southeast Neighbourhood Council had insufficient 'clout', expertise and experience to meet its responsibilities adequately. The co-ordination problems in the Bijlmer Sports Centre indicate the importance of properly defined authorisations and responsibilities for the parties involved in a disaster situation.

The fact that this disaster hit a relatively socially disadvantaged section of the city further complicated the work of the providers of relief and care. This was also an underlying factor in the abuse of provisions and benefits. At the same time, measures to withhold relief from those not properly entitled to it did give rise to considerable unrest among certain groupings.

The spread of victims across widely separated day care and overnight accommodation facilities created organisational problems and tension between the relief workers and their clients. The early centralisation of day care made it more manageable, since it enabled a better overview of the group of people affected and simplified co-ordination between the operational services involved.

Representatives of the victims were actively involved in the preparation and implementation of various measures. The city authorities sometimes encouraged this, as with preparations for the memorial service and the after-care plan. In other cases the stakeholders took the initiative themselves – as with attempts to limit distressing overflights of the Bijlmer. This type of participation in relief activities or in prevention of future tragedies can significantly help in the process of assimilating traumatic events.[34]

As it had not been covered in the city's disaster plan, the after-care plan had to be drafted at very short notice. It proved far from easy in a decentralised situation to manage and monitor the provision of after-care, imposing a relatively heavy task on the victims' social networks.

Notes

1 Psychological issues relating to the disaster, and the effects of after-care activities on the psychological well-being of those affected, were the subject of a study by the Psychiatric Department of the Academisch Medisch Centrum. See also M.B.R. Kroon and W.I.E. Overdijk (1993). Psychosocial care and shelter following the Bijlmermeer air disaster. *Crisis*, Vol. 14, No. 3, 117–125.; and B.P.R. Gersons and I.V.E. Carlier (1993), Plane crash intervention: A preliminary report from the Bijlmermeer, Amsterdam. *Crisis*, Vol, 14, 3, 109–116.
2 Specialist Bijlmer disaster after-care team interviewee.
3 Co-ordination Centre log book.
4 Co-ordination Centre interviewee.
5 Amsterdam Social Services Department, Evaluatieverslag (ramp)hulpverlening vliegtuigongeluk in de Bijlmermeer (Evaluation report).
6 Police interviewee.
7 Ghana Co-ordinating Committee (Ghana country team) interviewee.
8 Th. E. Drabek (1986). *Human System Responses to Disaster: An Inventory of Sociological Findings*, Springer, New York.
9 Amsterdam Social Services Department, Evaluatieverslag (ramp)hulpverlening vliegtuigongeluk in de Bijlmermeer (Evaluation report), 10.
10 Amsterdam Social Services Department, Evaluatieverslag (ramp)hulpverlening vliegtuigongeluk in de Bijlmermeer (Evaluation report), 13, 26.
11 B. Raphael (1986). *When Disaster Strikes: A Handbook for the Caring Professions*, Hutchinson, London.
12 Chapter 4 deals in more detail with the registration issue.
13 'Personal input important in hall', *Het Parool*, 19 October 1992. There are also indications that white victims in the Sports Centre felt rather uncomfortable. Suddenly they had become a minority among minorities. Amsterdam Medical Services Department interviewee.
14 The Co-ordination Centre log book does not show precisely when the decision was taken to designate the Southeast Neighbourhood Council as the body responsible in the Bijlmer Sports Centre. According to the police log book in the Bijlmer Sports Centre, the Mayor took this decision on Thursday 8 October. Within the Southeast Neighbourhood Council itself, there seems to have been some confusion on this point, as demonstrated by interviewees' failure to agree whether administrative responsibility was obtained as early as the first night or only in the second half of the first week.
15 Police interviewee; Southeast Neighbourhood Council interviewee.

16 There were 12 country teams covering Africans, Antilleans, Arabic speakers, Arubans, Ghanaians, Israelis/Jews, Cape Verde Islanders, Dutch, Pakistanis and Indians, Spanish speakers, Surinamers and Turks; Rapportage Nazorgteam GG & GD Vliegramp Bijlmermeer (Report of the Medical Services after-care team at the Bijlmer disaster), March 1993.
17 Bijlmer Sports Centre interviewee.
18 Police interviewee; Southeast Neighbourhood Council interviewee.
19 Amsterdam Social Services Department, Evaluatieverslag (ramp)hulpverlening vliegtuigongeluk in de Bijlmermeer (Evaluation report), 5.
20 Amsterdam Social Services Department Evaluatieverslag (ramp)hulpverlening vliegtuigongeluk in de Bijlmermeer (Evaluation report), 5; city Social and Medical Services Department interviewees.
21 Police interviewee.
22 The literal and figurative distances between City Hall and the Bijlmer Sports Centre is mentioned by interviewees from both the Co-ordination Centre and the Southeast Neighbourhood Council.
23 K. J. Tierney (1989). The social and community contexts of disaster, in R. Gist and B. Lubin (Eds), *Psychosocial Aspects of Disaster*, Wiley, New York; R.W. Perry and A.H. Mushkatel (1986). *Minority Citizens in Disasters*, University of Georgia Press, Athens.
24 For differing viewpoints on the purpose of the day care centre, see the Co-ordination Centre log book and the city Medical Services Department evaluation report, 11, 14. The run-down of relief in the Bijlmer Sports Centre is dealt with in Chapter 5.
25 Netherlands Antilles relief worker interviewee.
26 Amsterdam Medical Services Department interviewee.
27 Ghana Co-ordinating Committee (Ghana country team) interviewee.
28 J.S. Meijer (1992). Een vliegtuigramp in een huisartsenpraktijk: posttraumatische reacties in de eerste vier weken na de ramp in de Bijlmermeer. *Nederlands Tijdschrift voor Geneeskunde*, Vol. 136, 2553-8.; Rapportage Nazorgteam GG & GD Vliegramp Bijlmermeer (Report of Medical Services after-care team at the Bijlmer disaster), March 1993; Amsterdam Medical Services Department interviewee; Netherlands Antilles relief worker interviewee.
29 Rapportage Nazorgteam GG & GD Vliegramp Bijlmermeer (Report of the Medical Services after-care team at the Bijlmer disaster), March 1993; specialist Bijlmer disaster after-care team interviewee; Netherlands Antilles relief worker interviewee.
30 H. Lakmaker (1992). Nazorg in de Bijlmermeer, *Tijdschrift voor de Sociale Sector*, Vol. 12, 14–18.
31 City Medical Services Department Co-ordination Centre log book, 40, versus interview with Netherlands Antilles care worker. From the Antillean corner one hears that psycho-social after-care is mainly provided by regular 'white' regional psychiatric care agency personnel. Although the Antilleans suffered the highest death toll, the regional agencies are said to have refused the offer of a team of Antillean psychiatrists and psychologists.
32 P.E. Hodgkinson and M. Stewart (1991). *Coping with Catastrophe: A Handbook of Disaster Management*, Routledge, London.
33 Rapportage Nazorgteam GG & GD Vliegramp Bijlmermeer (Report of the Medical Services after-care team at the Bijlmer disaster), March 1993. The after-care team only visits those directly involved, i.e. people who have lost someone near, or whose home was destroyed. Some 'reachout' work has been done by the Bijlmer Air Disaster Task Group of the Amsterdam Southeast Churches Liaison Committee. However, the organisation did not represent all the relevant religions.
34 B. Raphael (1986), op. cit.

CHAPTER 5

In aid of survivors: material relief

Introduction

At a single stroke the crash of the Boeing 747 destroyed 31 flats in the 'Groeneveen' and 'Kruitberg' buildings. The inferno that followed affected a further 49 flats which were already more or less damaged by the crash. Within a matter of minutes, a large number of people who had survived the disaster were literally deprived of hearth and home. For the relief organisations this meant an extra task on top of providing medical and psycho-social support for those affected. Extensive tangible provisions would also have to be made for a lengthy period of time. This would also involve facilitating the efficient processing of insurance claims and correct channelling of funds being provided by El Al Israel Airlines, the Dutch government and the European Community. The city Social Services Department was one of the implementing agencies. A more complex and striking side of material relief was the provision of replacement homes, furnishings and fittings, and immediate benefit payments to a large but hard-to-define group of victims.

This chapter focuses on the latter aspect of material relief activities. The principal agencies were the city Social Services and Housing Departments. It should be remembered, however, that, at various times, many other authorities and services were also intensively involved in material relief issues. In this chapter we distinguish between the *supply* of and the *demand* for material provisions. While providing adequate relief resources, the authorities also had to arrive at a clear definition of the group of legitimate applicants (the 'real' victims) for material relief.

It is also useful to separate *policy-level* decision making by the authorities in and around the Co-ordination Centre, from *operational* implementation by municipal and other agency representatives on the spot (the Bijlmer Sports Centre, the hotels, the offices of the city services). Combining these two dimensions results in four analytical perspectives: supply/decision; supply/implementation; demand/decision; demand/implementation. These are dealt with consecutively in this chapter. Dissecting the episode in this manner enables us to unravel the complex and controversial chains of events around this issue of material relief. It also helps put some of the more intensely politicised events of the entire crisis episode into a broader perspective.

Determining supply: the authorities' perspective

Reconstruction of the policy-making process shows five strategic decisions that dictated the supply side of relief provisions. We will deal with these decisions in chronological order.

Decision point 1: co-ordination of material relief

The need for an extensive programme of material relief soon became clear to the Co-ordination Centre. Despite this, in the early stages the central authorities in the Co-ordination Centre focused on the recovery and identification of fatal casualties and the tracing of those flat residents who had been reported missing. There was a short discussion on material relief on the morning of Monday 5 October. It was decided that the city Social Services Department should be made responsible for the co-ordination of material relief. The decision was understandable enough in that the Social Services Department is involved in a wide range of material relief activities in the wake of smaller-scale calamities, like fires. This department was used to dealing with benefit payments, supply of household necessities and advice and assistance in coping with victims' specific financial and tangible problems. The decision was noted and accepted by the city Housing Department.

With benefit of hindsight, this initial co-ordination decision had some inadvertent consequences. The underlying reason for this was that the decision was not communicated clearly enough to all parties involved. It took up to 10 days before some Social Services executives realised that their department was also meant to be representing the city Housing Department in the Co-ordination Centre.[1] Moreover, the decision was made while the Alderman responsible for Social Services, Wildekamp, was still on her way back to Amsterdam from abroad. She was informed of the co-ordination decision by letter, but the actual implications of the decision remained unclear. Moreover, immediately on return, Alderman Wildekamp's own responsibilities demanded her full attention. Her portfolio also included the city Medical Services Department and she was actively involved in after-care planning. Hence, it was some days before she could turn to this extra task. In the first week after the disaster, co-ordination between Alderman Wildekamp and Alderman Genet (responsible for the Housing Department) was largely absent.

All parties involved underestimated the need for co-ordination between the

Social Services and Housing Departments, and other bodies and services. Right from the very start, the Housing Department concentrated on its primary task, and in this regard was working far more closely with the private landlords, i.e. the housing associations, than with other city services. Moreover, agency head Denijs assumed that, under the circumstances, one should fully accept decisions made in the Co-ordination Centre, even if the Housing Department was not actually represented there.

By the same token, the Social Services Department was also primarily involved with its own tasks. These were highly varied and required deployment of staff at many locations spread widely across the city. On top of this, during the first few days the city Social Services Department was mainly occupied in providing emergency assistance; and from the very start the Housing Department was working towards permanent solutions to the accommodation problem. The difference in focus meant that, in the Co-ordination Centre, the Social Services Department was a less than optimal representative for the Housing Department.

Decision point 2: dealing with illegal immigrants among the victims

During the night and morning of 5 October, the Co-ordination Centre started to receive reports from a number of sources indicating that illegals among the victims were too frightened to report in for relief and/or material assistance. Looking back, this was understandable and predictable. Previously, the group of people involved had always tried to stay out of sight of the authorities. And in the months preceding the air crash their resolve to remain invisible had been strengthened by the general hardening of Dutch public opinion on the inflow and presence of illegal immigrants. Put more directly, the illegals in the Bijlmer simply did not trust the city authorities, or the police in particular. Some months before, this feeling had been heightened by the arrest of a large number of suspected Ghanaian drug dealers; this in turn prompted public statements by the police to the effect that there were a very large number of illegal Ghanaians in the neighbourhood, many of whom were engaged in criminal activities.

The crash posed additional problems. Illegals, whether injured or not, traumatised and deprived of their homes, feared that reporting to a reception centre could eventually cause them to be expelled. So they moved in with friends or wandered around. On hearing about their precarious situation, Mayor Van Thijn contacted the embassies of some of the countries involved (including Ghana), as well as relief workers and organisations representing illegal immigrants. Information from these sources convinced him of the need for swift action. Urgent reports on the problems also came from the police.

The Mayor duly decided to make an appeal to all the illegals affected, saying that they had nothing to fear and that they should report to the various reception centres. He guaranteed that there would be equal treatment for all, irrespective of immigration status. Essentially, the Mayor's statement meant that there was no risk of these 'illegal victims' getting expelled. At the same time it did not mean that their residence status would be legalised.

There were two reasons for the decision. On humanitarian grounds it was regarded as unacceptable that illegals who had suffered the effects of the disaster should be deprived of material and other relief, which was freely available to other victims. Moreover, the registration and co-operation of the illegal flat-dwellers and their friends and relations were crucial to the police investigation into the number of victims and the fate of the missing persons. The Mayor was most unhappy with the idea that a lack of information from illegal circles might prevent a possibly large number of the fatalities from ever being identified, whereby they would have died anonymously and unnoticed.

The Mayor's decision was supported in the Co-ordination Centre. Strictly speaking, policy decisions with regard to illegal immigrants fall outside the normal authority of the Mayor. By law the Chief of Police is head of the local Immigration Department and as such is directly responsible to the Minister of Justice. However, Mayor Van Thijn and Chief of Police Nordholt were in close agreement on the course to be followed, allowing no room for considerations of competency to interfere in their working relationship.

The dominant perception among illegal immigrants was that the appeal was mainly designed to aid the police investigation, and that as soon as the disaster was over the police would certainly use the information gained to their detriment.[2] Implementation of the decision required a dual standard, even a paradoxical approach by the authorities: the left hand was not allowed to know what the right had learned. The police were put in a particularly tricky situation. The investigating team was wrestling with an enormous number of missing persons reports, and was constantly under heavy pressure to meet the X-Hour deadline; naturally they were most anxious for information from the illegals. At the same time the decision meant that the officers involved – despite all their training and usual police practice – were not to inform their colleagues in the Immigration Department on illegal residents. It was this latter department that, under heavy public and political pressure, had the task of reducing the number of illegal immigrants in the Amsterdam area. The situation was just as peculiar for officials of the Social Services and Housing Departments, and the City Population Register.

Amsterdam's initiative prompted two reactions from the State Secretary for Justice. On 7 October, he wrote to the Mayor and the Chief of Police. In his letter to Van Thijn, he emphatically endorsed the decision. However, his letter to Chief of Police Nordholt contained a significant proviso: 'Where the person concerned has been here many years, without a break, and is in permanent employment with social security premiums and taxes duly paid, and where no further objections exist, a residence permit *may* be granted on humanitarian grounds. *Where these conditions are not met, the person concerned will have to leave the country.*'[3] At the time, the authorities in Amsterdam accepted this for reference.

The Mayor and the Chief of Police both maintained the position that at this point Amsterdam should act as it judged best, and that the guarantees given should be honoured in full.

Decision point 3: scaling down relief in the Sports Centre

Developments in the Bijlmer Sports Centre are dealt with in detail in Chapter 4. Decisions on the situation there were dealt with in the Co-ordination Centre. At the end of the first week, a plan was drafted that covered a rapid scaling down of provisions in the Bijlmer Sports Centre. On Wednesday 7 October the agencies providing material relief had received the impression that their work in the Sports Centre had almost come to an end. They believed that they had a clear picture of who was and who was not entitled to replacement housing.

Moreover, the Sports Centre appeared to exert a powerful attraction. Although the many victims among the flat-dwellers had lost no time in applying for replacement homes, there was no appreciable fall in the number of people calling at the counters of the Housing and Social Services Departments in the Sports Centre's main hall. In addition, the atmosphere in the hall was deteriorating. On Thursday 8 October, Director Denijs of the Housing Department heard from his staff that department officials were subject to increasing aggression from visitors. They also had the impression that hardly any new 'genuine' victims were reporting in.[4] Their colleagues at Social Services agreed. On Friday 9 October, Alderman Genet consulted with the Social Services, Housing and Medical Services Departments and the police on the situation. It was duly agreed that the centre would be closed on the following Monday or Tuesday.

Attention and effort in the Co-ordination Centre that Friday had been focused on organising the release of the list of dead and missing. Preparations for the communal service of remembrance and a possible collective funeral also made heavy demands on time. The planned closure of the Bijlmer Sports Centre went virtually unnoticed in the Co-ordination Centre.[5] However, over the weekend, opposition to the scheduled closure grew. The Southeast Neighbourhood Council had not been present at Genet's meeting with the city departments and the police. That was remarkable in itself given that the Co-ordination Centre had designated the Southeast Neighbourhood Council as the body responsible for the sports hall. This demonstrated the tension between the various co-ordination competencies: operating from the centre, the Social Services Department co-ordinated the provision of material relief in the broad sense; the 'de-central' Southeast Neighbourhood Council co-ordinated the activities of all services for both material and psycho-social relief located in the Bijlmer Sports Centre (at least according to the Co-ordination Centre). Representatives of the Southeast Neighbourhood Council claimed it was too simplistic to suppose that the continual inflow to the Sports Centre was solely due to 'free riders'. Victims could be reporting in to the responsible organisations only now, because they had been recovering from the initial shock for a couple of days with family or friends.[6] It was further argued that the Sports Centre had acquired an important symbolic function in the local community. It had become an important link in the collective mourning and assimilation process, and could not be removed just like that.[7]

On the morning of Monday 12 October, the Mayor proved sympathetic to these arguments. The fact that the Southeast Neighbourhood Council made a serious issue of it also played a role. In many ways, the Southeast Neighbourhood Council had been moved into a subordinate position in which its actual function was as an implementer of policy decisions taken at City Hall. Some neighbourhood councillors felt that it was quite unacceptable to be ignored yet again by the central authorities and city agencies on this particular point; moreover, they maintained that there was still a very real need for relief and care in the Bijlmer Sports Centre.

The Co-ordination Centre reversed the original decision to scale down use of the Bijlmer Sports Centre, and decided that it should stay open for at least one more week. On Monday morning, the Head of the Housing Department was informed that they were expected to maintain a full and undiminished presence in the hall. The reversed decision met with strong disagreement. Eventually, the issue was more or less settled by the intervention of Alderman Wildekamp; as the Alderman for Social Services, she had final responsibility for the co-ordination of material relief. A compromise was reached.[8]

However, the issue was not settled yet. On Thursday 15 October, it became clear that Mayor Van Thijn wanted the Housing Department and other services to maintain a presence in the Bijlmer Sports Centre until it was closed down altogether.[9] Director Denijs of the Housing Department received this news on his cellular phone while making an inspection visit.[10] But Denijs' staff had already told him, in no uncertain terms, that they were against a longer stay in the Sports Centre, saying that it was unnecessary from the angle of actually doing the work, and that they felt physically threatened. Denijs sided with his staff and twice declined to obey a formal instruction to send his officials to the Sports Centre. Finally, Denijs was given an official order: further refusal to co-operate could have 'grave consequences' for his position.[11] Thus confronted, Denijs abided by the formal decision. At the same time, it was then decided that the Housing Department should have its own representative in the Co-ordination Centre, so ending the unsatisfactory representation by the Social Services Department.

Decision point 4: towards a residency arrangement for 'illegal' victims

The decision to treat all victims the same, irrespective of their legal or illegal immigration status, only produced to a limited degree the clarity and harmony envisaged. On Tuesday 13 October, the full risks and difficulties of implementation were graphically highlighted. Six illegal Ghanaians had reported in, and applied for relief. Social Services referred them on to the Police Immigration Department where – contrary to all the earlier public guarantees – they were duly registered. The news spread like wildfire through the Bijlmer.

The news caused anger and frustration among members of the Co-ordination Centre. At that evening's regular press conference, attempts were made to clear the air by announcing that a mistake had been made and that the relevant files would be destroyed immediately. Under the circumstances, it was an adequate attempt at damage limitation. But the damage had already been done: unrest spread fast among the illegals, and the authorities in the Co-ordination Centre were criticised from all sides. The press seized on the issue. The climate of public opinion began to harden. Significantly, the press conference on Wednesday morning, 14 October was interrupted by a bomb scare originating from a group calling itself 'RaRa' (a well-known radical group responsible for several semi-terrorist incidents), which claimed that this was a protest at the way City Hall was treating the illegals.

There was a feeling in the Co-ordination Centre that something must be done to achieve a turnaround. Chief of Police Nordholt in particular was of the opinion that the logical consequences would have to be faced, namely that the illegal victims concerned would have to be given residence permits. To deprive them of this would inevitably lead to more incidents. The issue was discussed by a small group in the Co-ordination Centre. Strictly speaking, the Chief of Police was responsible for the implementation of immigration policy. His feeling was that the initial decision on equal treatment of illegals had placed him in a very tricky situation, both *vis-à-vis* his own Immigration Department and public opinion.[12] He wanted to get matters back on a sound footing and to arrive at an *ad hoc* arrangement for the legalisation of the illegal immigrants affected by the disaster, in direct consultation with the Ministry of Justice in The Hague. This would better achieve the Co-ordination Centre's original, humanitarian aims. The Mayor concurred.[13]

Nordholt consulted with State Secretary of Justice Kosto. It was agreed that illegals who were victims of the disaster would be eligible for legalisation. A press conference was called to announce the arrangement. The list of candidates for legalisation was to be drafted under the supervision of Alderman Wildekamp and then submitted to the Mayor and Justice Minister for a formal decision. Nordholt's public performance in The Hague took the members of the Co-ordination Centre by surprise. They were unaware of the fact that Kosto and Nordholt were to give a joint press conference.[14] Possibly even more relevant was that Alderman Wildekamp had not been informed of her new task in advance. So, Alderman Wildekamp was suddenly confronted with a new, politically and socially highly sensitive, dimension to the registration activities already underway.[15]

Decision point 5: continuation of relief and services in the hotels

By the end of the first week it was clear that the separation of day and overnight reception and relief facilities had achieved its purpose, and that it was time for a new arrangement. Moreover, several of the officials responsible for psycho-social care (see above) feared that the Bijlmer Sports Centre might be turning into some sort of quasi-permanent 'temple'.[16] Moreover, the naval barracks were no longer available for overnight accommodation. To put it succinctly, new accommodation would have to be found. The housing of Antillean and Aruban victims in a hotel, paid for by the Netherlands Antilles, meant that there was little choice other than to house the other victims in an equivalent fashion, i.e. also in a hotel. Hence, the Mayor instructed the Social Services that the other clients for emergency accommodation should also be housed in an hotel.

Co-ordinating supply: implementation

Housing Department and Social Services staff showed a high degree of commitment even when working under heavy pressure. The Housing Department soon developed an excellent working relationship with the Nieuw Amsterdam Housing Co-operative and was able to locate sufficient replacement homes within a very short time. The Social Services Department confronted a wide spectrum of differing tasks, ranging from benefit payments to help with rehousing, to arranging transportation between reception centres, and to taking care of funeral details. A crisis centre was established at the Social Services headquarters to provide co-ordination. Even so, it proved difficult for the various units within Social Services to get a full overview of what the others were doing and to co-ordinate activities. However, this did not pose any insurmountable obstacle to this department's primary services, and in most cases matters proceeded satisfactorily. The main implementation problems in establishing the supply side of material relief both involved co-ordination.

Where opposites meet: Social Services and Housing

Co-ordination and liaison between the Social Services and Housing Departments was difficult. At an early stage, in the Co-ordination Centre, the Social Services Department had been designated as co-ordinator for material relief, but was unable to do this adequately (see decision point 1 above). The problem was most visible in a series of communication breakdowns, causing repeated confusion and irritation. Social Services were not sufficiently familiar with the basic working processes of the Housing Department, and vice versa. And so, on Thursday 8 October, the Housing Department and the Nieuw Amsterdam Housing Co-operative were somewhat surprised when representatives of the Social Services contacted them 'to see about arranging for replacement homes', whereas in fact they were already well advanced with supplying new housing for the people who had lost their flats.[17]

From the very start, the two city departments adopted differing approaches to the issue. Whereas at Social Services the Special Projects Unit was given an important role in arranging for relief, the Housing Department used as much as possible its usual *modus operandi*. In practice, a branch office was cleared for disaster activities, while regular clients were dealt with at other branches. While the Social Services formed a project organisation, including a crisis centre under the head of the City Operations Unit, the Housing Department's normal hierarchical line stayed largely intact. This did not help to promote clarity in mutual co-ordination.

Policy versus implementation: the illusion of co-ordination

Another problem related to the communication between the decision-making level in the Co-ordination Centre and the operational level (see also Chapter 6). This is evident from a number of incidents. As noted earlier, the people at Housing believed that Social Services were representing them in the Co-ordination Centre. Hence, they also assumed that information from the Housing Department would reach the Co-ordination Centre via Social Services, and that outgoing instructions would follow the same channel. As the Housing Department heard nothing from Social Services in the first few days, its Director Denijs and his staff assumed that the Co-ordination Centre was satisfied with their performance. As it happens this was the case, but the Co-ordination Centre had made a number of enquiries and remarks in the direction of the Housing Department, and these had not come through. On its own initiative, the Housing Department was doing more or less what the Co-ordination Centre wanted. This created the impression in the Co-ordination Centre that the Housing Department was well integrated into the system of reports, instructions and feedback.

This mutual 'illusion of co-ordination' only came to light at the time of the conflict over the scaling down of the Bijlmer Sports Centre. At the Co-ordination Centre, Alderman Genet was also unaware that there really was no co-ordination of information flows between City Hall and 'his' Housing Department. Neither could he have known that the flow was actually blocked by Social Services, nor indeed of the remarkable situation in which the Housing Department was operating more or less autonomously while assuming that it acted with the explicit knowledge and approval of the Co-ordination Centre.[18]

The distance between relief policies and implementation was further increased by the highly specialist, technical approach that seemed to pervade all levels of the Housing Department. Clearly, it managed business within its own territory extremely well. However, the sense of being part of a much larger operation was less well developed. This was even more sharply relevant in a socially and politically sensitive context, such as this, where policy perspectives may dictate a less rigid operational stance. The Director and his staff found it difficult to accept that more was involved in the scaling down of the Sports Centre facilities than the efficient processing of housing matters and their working conditions and equanimity. Here again, the physical distance between the Co-ordination Centre and the Housing Department played a role: a presence at City Hall would have made it easier to comprehend the need for a broader outlook.

Within city Social Services itself there was a different problem with regard to co-ordination of policy and implementation. Distortions and misunderstandings clogged information flows between units in the field, their own crisis centre, their representatives in the Co-ordination Centre at City Hall, and the Co-ordination Centre as a whole. Initially this was a technical communications problem.[19] But as the disaster continued and the Social Services operation came under ever-increasing pressure, it became clear that it really involved an organisational problem. Information from the operational level and from various niches penetrated too rapidly and without verification into the Co-ordination Centre. There was no management level to check and correctly distribute the incoming information in the Co-ordination Centre.

Regulating demand: the authorities' perspective

We have reconstructed the decision-making process in the Co-ordination Centre in three phases. We have quite deliberately avoided the term decision point here because, at the Co-ordination Centre level, remarkably few explicit decisions were taken on this issue.

Phase 1: maximum relief

During the first few days, there was hardly any explicit discussion in the Co-ordination Centre on the need to take measures to obtain and to maintain a clear picture of the group of victims. This phase chiefly concerned the identities of the fatal casualties and uncertainty on the fate of the many missing persons. The Co-ordination Centre did not get round to setting more-or-less formal limits on provision of relief. Moreover, in the first few days there was a strong urge to provide 'as much help to as many as possible'. The Mayor's explicit wish not to differentiate between legal and illegal victims set the tone in this regard. Item number one was to ensure an adequate supply side. On top of this, in the eyes of the authorities in the Co-ordination Centre, neither effort nor expense should be spared; they knew that they would be covered by promised financial support from the Dutch government and the European Community. Moreover, most members of the Co-ordination Centre clearly did not expect critical questions about some benefitting at other people's expense, certainly during the immediate disaster period. And later on it would surely be politically inopportune for council members or parliamentarians to be portrayed as taking the position of stringent accountants, against the background of a major tragedy like that in the Bijlmer. The word from the Co-ordination Centre was – to paraphrase – 'better to help too many than risk depriving any single victim'. It was assumed in the Co-ordination Centre that relevant services would apply their own standards and procedures for determining eligibility, whether or not in consultation with the responsible Aldermen.

Initially, the implementing departments had no difficulty with the latitude given them. They worked according to their own tried and tested methods. However, there were very clear differences in working practices between the Housing and Social Services Departments, and these were not always easy for claimants to appreciate. Typically the necessity, type and formality of documentation needed varied from one department to the other, although they sometimes sat side by side, like brothers, in the Sports Centre.

After just three days, officials of the various departments were starting to report that more and more people appeared to have been attracted by what was going on. The Housing Department soon had what they believed was a good picture of the group of victims. Hence, anyone who subsequently reported in claiming to be a victim was very likely not to be entitled to help. However, these impressions were not conveyed clearly enough to the Co-ordination Centre. Social Services personnel at the sports hall were fast starting to be convinced of a growing number of 'free riders'.

Being suspicious of fraud, the departments involved adopted a more cautious stance. As the week wore on they became more and more convinced of the need for more stringent procedures and meaningful verification of claims. The departments for years had tried to shield their clients from City Hall's pressure for uniformity and spending cuts. But now, in this extraordinary crisis situation, they found it hard to relinquish the business-like style – which had also by now penetrated their ranks – and to apply the soft pedal when interpreting the rules. For them it was sometimes hard to absorb that this time it was City Hall that was saying that the emergency justified and required a temporary departure from the normally tough restrictions on spending public funds.

Phase 2: towards the Kosto list

Despite the authorities' philosophy of maximum solidarity, it was becoming ever clearer that a halt had to be made to the continuing large number of people using the day and overnight facilities. When the remaining victims were moved to the

Novotel, a screening procedure was used. Previous attempts at screening, in particular the pass system, had failed (see Chapter 3). But once again during the transfer to the Novotel it proved difficult to establish and maintain unambiguous criteria. All agencies involved, adopting different criteria, had developed their own lists of so-called 'entitled victims'.

The existence of different lists of 'entitled victims' prompted a comparison of the records of the Social Services and Housing Departments, the Nieuw Amsterdam Housing Co-operative, the City Population Register, and the police. The aim was to arrive at a more or less definitive list of victims entitled to the provisions being offered. This list helped exclude those not properly entitled from further relief and material assistance.

Once again, this initiative does not appear to have been the outcome of explicit decisions in the Co-ordination Centre. It was much more a matter of Alderman Wildekamp exercising her authority on the grounds of reports from the departments operating under her responsibility. Even so, the initiative was supported in the Co-ordination Centre, and instructions came from there to the effect that people staying in the hotels should be screened as soon as possible.

Comparison of the various lists started on Wednesday 14 October. For the second time in the disaster period, considerable energy had to be devoted to compiling and checking lists. Happily, in this case it involved less harrowing categories than the dead and missing dealt with the week before. The possible legalisation of illegal immigrants who were recognized as victims gave greater weight to the initiative to compile a new list – and to the irretrievably intertwined social and political ramifications. It required the bodies concerned to formulate the criteria both stringently and fairly.

Phase 3: towards 'objectivity'

The compilation of the final list prompted considerable discussion in the Co-ordination Centre. Chief of Police Nordholt insisted that the police should in no way be involved in the implementation of the agreement with Kosto. This would rule out the use of the now well-practised team of police investigators. At the same time, registration and verification of the personal details of the people in the two hotels was creating its share of problems. Social Services staff encountered problems of communication and, on occasion, even aggression. The members of the Co-ordination Centre also realised that there was a possibility that not all legitimate candidates for the Kosto list were lodged in the hotels. Some might well have gone into hiding with family or friends. This group should also be given the chance to take up the State Secretary's offer, but this would require them to report to the city authorities.

This eventually led to two decisions in the Co-ordination Centre. Firstly, the people lodged in the hotels would, after all, have to be re-registered. Secondly, an appeal would be issued via the media to all illegal victims who had not yet reported in, to do so now. The staff of the City Population Register were entrusted with registration for the Kosto list; they were to be supported in the hotels by the Social Services Department. The Southeast Neighbourhood Council was given the job of re-registration. The Co-ordination Centre insisted that procedures for inclusion in the Kosto list should be made as 'objective' as possible. In other words, candidates must show incontrovertible proof – i.e. preferably by some sort of written document of legal status – that they were living in the flats at the time of the crash.

This approach would avoid the registration procedure turning into an extensive investigation to verify the many personal stories and declarations by witnesses submitted by candidates for the Kosto list. Making matters objective would simplify the final decision making on who was eligible. Even so, Alderman Wildekamp and the Director of the City Population Register, Geuzinge, were not to be envied. Against the traumatising background of the Bijlmer disaster they had the thankless duty of facilitating the inevitable but inherently controversial switch from indiscriminate emergency relief to rigorously selective, structural care and assistance.

Geuzinge would later call it an 'impossible task': one would have to disappoint large groups of people and every mistake or apparent mistake would be a blow to morale and public opinion.[20] An additional factor here was that Geuzinge and his department were on the receiving end of an increasingly critical press. The long drawn-out uncertainty as to the large number of missing persons and now the ups-and-downs of registration of the victims among the illegal flat-dwellers prompted critical press coverage of the City Population Register's performance: 'Government loses overview of citizens: Population register not cut out for disaster'.[21] The Register's records were said to be largely inaccurate. The picture was of a dusty bureaucracy with little grip on its core task. In this climate, Geuzinge's careful explanation of the City Population Register's *modus operandi* and the almost insoluble issue of registration with regard to the official residence of sub-tenants and illegal immigrants was of little help.

Controlling demand: implementation

Policy decisions on the provision of help and assistance worked through to the demand side. Initially the implementing departments worked separately. However, it became increasingly apparent that there would have to be a more co-ordinated approach to identification of the list of clients. Decision making on the Kosto list fostered a more formal definition of the problem. We will concentrate on two themes that caused much commotion during the disaster period: the deteriorating relations between relief workers and their clients, and the mass assault on the offices of the City Population Register.

From solidarity to constraints

Direct contact with the public is routine for both the Housing Department and Social Services. As street-level services, not only do they know the area of work and the clients, but they also come into contact with people trying to obtain benefits to which they are not entitled.[22] Over the course of years there has been a steady formalisation of procedures and regulations to safeguard the fair and careful allocation of city benefits and provisions. During the Bijlmer disaster it was very clear that the city services were heavily influenced by their organisational culture when setting up the material relief programme. Although fully aware of the need to provide maximum help to the victims, both departments still appeared unwilling to relinquish their usual meticulousness.[23]

Finding a balance between the two hard-to-reconcile demands was far from easy. An inevitable consequence of differentiation was that somewhere, somehow the bad news had to be told to those people who were not entitled to help under the criteria. In some cases, rejection provoked aggression – not least because it implied in fact a formal denial of victim status. In a few cases this actually led to scuffles between Social Services staff and people who claimed to be victims. Added to this, the immigrant grouping concerned already felt generally disadvantaged for a variety of reasons. Moreover, the rejections were conveyed in the understandably emotional and tense atmosphere of the sports hall. And so, the daily selection of who was and was not to be sent to the emergency accommodation developed into a hazardous and loaded undertaking, with rage and frustration soon bubbling to the surface.

Towards the end of the week the contrasts between the business-like approach of the city departments and the more emotional demand for solidarity and support from many different people became even stronger. Increasingly, the relief services were turning people down because they were not classified as victims. This problem took on even more complex aspects during the time at the hotels. The failed screening process before people were moved to the Novotel, and the fact that the Antillean group had been moved to Park Lake Hotel completely outside the auspices of the city authorities, made it necessary to check people's status after the move. This introduced additional sensitivities into the situation: initially people were admitted to the two hotels as victims, but after evaluation they could lose both this status and access to the assistance and benefits from the city. Against this background, it is hardly surprising that some aggression was directed at the Social Services officials conducting the registration.

Attempts to separate the entitled from the non-entitled persons in the Park Lake Hotel led to a fairly critical situation. Despite earlier agreements, the FORSA Foundation, the Antillean-interests organisation that was actually in charge at the Park Lake Hotel, declined to help city officials in checking their people. Responding to reports from their own organisation, Social Services representatives in the

Co-ordination Centre were unwilling to have their staff exposed to aggression from people accommodated in the Park Lake Hotel.

This stalemate led to heated discussions on 16 October and an emergency meeting of all parties involved on Saturday 17 October. In a long session Deputy Mayor De Grave sought carefully to mediate between the Social Services Department and FORSA, and to deal with the problems that had arisen. This required cautious navigating by City Hall: on one hand due respect for rules and criteria for victims, and on the other the weighing of Antillean interests so that FORSA would not be driven into a corner.

The attempts at mediation eventually produced a satisfactory result: FORSA promised to co-operate in the separation of those who were or were not entitled to relief, and would continue to manage the operation in the hotel. Social Services agreed after all to the deployment of its special control staff for the screening of people in the Park Lake Hotel. The actual screening took place on Monday 19 October.

Crisis symbolics: queues along the canal

The public appeal for the still-unregistered illegals to report to the Population Register did cause problems. The press conference given by the State Secretary and the Chief of Police provoked a major flow of people to the City Population Register desk in the Bijlmer Sports Centre. So many people were turning up on the afternoon of Saturday 17 October that it was decided on the spot to prolong the original deadline for closure of the Kosto list, and to ask people still waiting their turn to report to the head office on Monday morning. The reason was quite simply that the officials on duty could not process the volume at that time.

The extension of the registration deadline was a remarkable example of an operational decision with unforeseen strategic ramifications. In addition to everyone who had unsuccessfully applied at the Bijlmer Sports Centre on Saturday (designated as category I), the extended deadline for registration also applied to other victims who might, or might not, have been referred by other relief agencies to the City Population Register (category II).[24] The announcement of the extension – plus the start of rumours among the illegals of a golden opportunity for legalisation – further boosted the numbers applying. By the time all of category I had been processed on Wednesday 21 October, the City Population Register still had long lines of category II applicants queuing in front of its doors on the Herengracht, one of Amsterdam's most famous canals. This prompted the Director of the City Population Register to extend the deadline again for this group. Up to Friday 23 October, the day on which registration at the City Population Register was finally ended, almost 1,800 illegals found their way to its offices on the Herengracht.

Extension of the deadline also unintentionally increased the scope for 'free riders'. Rumours in the illegal community had their effect after Saturday 17 October. Press and television showed long lines of people waiting on the Herengracht. On Tuesday and Wednesday it was becoming quite clear that the vast majority of these illegals would not be eligible for victim status. Indeed, people driving cars with German licence plates also appear to have come to Amsterdam to try their luck. The view of the queues waiting to get into the City Population Register, opposite his official residence, was an unwelcome surprise for the Mayor.

The commotion surrounding the registration of illegal victims helped further to harden public opinion and erode the initial social solidarity with victims of the Bijlmer disaster. In turn, this prompted media criticism and a cynical public attitude toward illegals. The thought that there might well be genuine victims among the illegals reporting in gradually disappeared into the background.

For the City Population Register, compiling the Kosto list proved to be a heavy task with a long shadow to match. For months following the Bijlmer disaster, criticism of the workings of the City Population Register continued, but its eventual final selection still held up in court.[25]

Conclusions

The supply side of material relief can be recapitulated in the following key points:

- The Housing Department was a crucial body in providing material relief, and its initial absence from the Co-ordination Centre was counter-productive.
- Allocation of an external co-ordination task to an organisation – in this case the Social Services Department – which was already under extreme pressure in coping with its own, many-faceted and geographically spread, tasks, was a more-or-less spontaneous decision. The decision was taken somewhat over-hastily. It certainly caused a number of misunderstandings and communication breakdowns in the links between the Co-ordination Centre and the implementation level.
- The course of decision making on equal treatment for everyone irrespective of immigration status provides somewhat bitter proof of the cliché that good intentions are not enough. There is no doubting that the decision making always put the victims' interests first, but the members of the Co-ordination Centre consistently failed fully to appreciate the feasibility of carrying out these decisions in practice. The paradoxical aspects of the initial decisions were insufficiently discussed in the Co-ordination Centre. The possible attraction exerted by the legalisation decision was never a topic for discussion. The initiator of the legalisation, Chief of Police Nordholt, appeared convinced that

under these circumstances, there was only a limited chance of abuse and 'free riding' by illegal immigrants.[26]

- Reconstruction of decision making in regard to the run-down of the sports hall activities raises questions about the way city services and sections of the Neighbourhood Council organisation can be satisfactorily interfaced in a disaster situation. The position of Neighbourhood Councils in disaster situations needs to be articulated more clearly, which implies an assessment of the relative merits of centralised and decentralised forms of crisis response.
- Further, this episode clearly demonstrates the differing philosophies regarding the organisation of help at City Hall and at the various services: at moments of decision the Co-ordination Centre showed a bias in favour of linking material and psycho-social care, while departments involved in material relief followed another line. Both the Social Services and the Housing Departments saw the disaster primarily as 'more of the same', i.e. chiefly a quantitative problem. Hence, they went ahead according to their own conceptions of task management, and in due course this meant that they were ahead of some other departments in wanting to end the integrated supply of material relief. Their efforts to act along routine lines and deal with matters at their own locations clashed with the Co-ordination Centre's dominant idea of integral services.
- Lastly, decision making on the winding down of activities in the Sports Centre was notable for the criss-crossing of the usual portfolios and *ad hoc* agreements on various Aldermen's areas of responsibility. At the same time, while this did not result in any serious mutual problems, for some directors of city departments it did create uncertainty as to who precisely was responsible for what.

In turn, the reconstruction of the strategic and operational sides of the management of the demand for material relief invites another set of concluding observations:

- Throughout the disaster period the Co-ordination Centre was not as stringent in managing the demand as in deciding on the supply of material relief. Initially the Co-ordination Centre focused exclusively on establishing maximum solidarity. In the second week, when the problems of 'free riders' and inaccurate records became more urgent, considerable attention was devoted to restructuring the supply of provisions. This meant criteria development was mainly bottom-up and absorbed a relatively large amount of time and effort.
- In a number of ways the authorities underestimated the social dynamic provoked by the disaster, and subsequent measures in the Bijlmer. It is quite true that the emotional peaks at X-Hour and the collective mourning on Sunday 11 October were followed by the tumultuous switch from overnight reception

centres to the hotels, and the decision on possible legalisation but, to many officials, the operation still seemed to be more or less on course at that time. Several departments were getting ready for a return to the normal pattern of services. Nobody seemed to have taken account of the apparently major effect of opening the Kosto list to illegals in and around the Bijlmer. By the end of the second week, many seemed to feel that the end was in sight. They expected a down-scale, not another test of strength.

- This approach and the screening of a large number of people from varying ethnic backgrounds was a highly delicate problem for the implementing departments. Aggression directed at the Social Services officials prompted a hectic emergency meeting in the Co-ordination Centre. The employment of riot control procedures was seriously discussed. Police objections and the ensuing mediation by the Deputy Mayor eventually achieved de-escalation.
- The following chain of events illustrates how delicate the link can be between strategic and operational levels (i.e. policy and implementation) in crisis situations: the legalisation decision causing the allocation of registration tasks to the City Population Register; the appeal to as yet unregistered illegal victims to report in, causing the influx at the Sports Centre; the *ad hoc* decision to extend the deadline for registration of illegal victims triggering the five-day 'storming' of the offices of the City Population Register and the subsequent turnaround in public opinion on the disaster and the Bijlmer in general. This is a chain of largely unintended and unforeseen effects. And the run of events had negative ramifications. The question remains as to how far these unintended effects were foreseeable and preventable.

The actual work of registration and selection to determine demand for material provisions ultimately proved extremely difficult. Indeed, this may well have been the most problematic aspect of crisis management for the city of Amsterdam. It was the result of the unexpected scope and complexity of the housing and registration issue itself. In general terms the authorities were aware that the Bijlmer has a very heterogeneous population with a broad range of social problems. But the specific difficulties and the inaccessible sub-cultures of the various groupings of illegals had been invisible to many people until the moment of the disaster. Previously, signals to this effect from the side of the police appear to have had trouble getting through properly. The disaster seems to have helped make the issues more visible.

The difficulty in carrying out this unexpected, and socially sensitive, task was exacerbated by organisational problems. Getting through to people whose daily concern was precisely to stay well out of view of the relevant bodies was no sinecure. The formalisation and tightening up of the approach to users of the relief facilities – notably by officials of the Social Services and Housing Departments – encountered major opposition, first in the Sports Centre and later in the Park Lake Hotel in particular. The police instigated what they regarded as a necessary development, only to stay firmly in the background when it came to implementation. The authorities had not counted on the social dynamic set in motion when they offered a very wide category of people, with few or no real prospects in normal circumstances, the chance to better their lot. The necessary comparison and linkage of various listings and records also complicated the information flows between the police, the City Population Register, the Social Services and Housing Departments, the Nieuw Amsterdam Housing Co-operative and the Southeast Neighbourhood Council. During this phase the Co-ordination Centre appeared to have less grip on events, and on several occasions was forced into more of a reactive (i.e. the solution of a variety of mini-crises) rather than a directive role.

Many of the problems that arose at the material relief level were the outcome of the difficult interactions between supply and demand and policy and implementation. Given the very tight schedule and the fact that material relief had to be established under the spotlight of public opinion, decisions by the authorities could, and did, lead to unintended effects faster and more visibly than in a routine situation. At the same time, ostensibly innocent misunderstandings and communication problems soon caused an uproar at the operational level, and duly rebounded as a new set of problems flared up at the policy level. The fact that these did not grow out of all proportion demonstrates that City Hall's basic policy precepts could stand the test of fire.

The pros and cons of registration notwithstanding, the fact remains that several hundred people were offered a wide range of financial and material help within a relatively short time frame. Confronted with the unexpectedly large, complicated and partly self-imposed problems of implementation, the city's apparatus may have stumbled a few times, but it did show the necessary resilience. The fact that, in the final analysis, this was not enough to halt the erosion of social solidarity with the Bijlmer and its residents is an immutable legacy of the disaster.

Notes

1 Amsterdam Social Services Department interviewee; Amsterdam Social Services Department, Evaluatieverslag (Ramp)hulpverlening Vliegtuigongeluk in de Bijlmermeer (Evaluation report).
2 Compare with *Het NRC Handelsblad*, 2 February 1993.
3 Letter from A. Kosto to E.E. Nordholt, dated 7 October 1992, in Co-ordination Centre log book, Annex 49 (our italics).
4 H. Denijs (Director City Housing Department), *Vliegramp in de Bijlmermeer* (Air Crash in the Bijlmermeer; personal diary).
5 Co-ordination Centre log book, Friday 9 October, 15.25–15.40; but also see the first signs of disagreement on this subject at briefing 23 at 20.34, whereby the Southeast Neighbourhood Council was already pressing for a delay and referred to the City Medical Services' plans in this regard.
6 Southeast Neighbourhood Council interviewee.
7 Academic Medical Centre interviewee.
8 Housing Department interviewees; H. Denijs, op. cit.; Co-ordination Centre interviewee.

9 H. Denijs, op. cit., 25–26. The Co-ordination Centre log book notes this incorrectly as having occurred on Tuesday 13 October (124).
10 Denijs, op. cit..
11 Denijs, op. cit.; Co-ordination Centre interviewees.
12 Another interpretation, also seen in the press, is that the initial decision by the Mayor put the Chief of Police in a difficult situation, in that – in a normal situation the Mayor has no authority whatsoever in this area; hence the legality of the decision could be brought into question: 'Nordholt and Kosto save Van Thijn's good name', *Het Parool*, 16 October 1992.
13 Recollections and interpretations by the various members of the Co-ordination Centre differ somewhat on this point, particularly regarding to what degree and by whom there was discussion about the implementation of the decision, and its possible ramifications, at this juncture. However, there is no doubt as to the leading role played by the Chief of Police in arriving at the decision, nor is it doubted that the Mayor had given agreement for him to go to the State Secretary, although, of course, this was not a formal requirement.
14 Co-ordination Centre interviewee; moreover, Nordholt's presence at the press conference at the Justice Ministry appears to have been more or less accidental. After their meeting, the State Secretary asked Nordholt to accompany him to the press room.
15 One plausible interpretation is that the Chief of Police was aware that she was preparing a list and used the opportunity of ensuring adequate registration, without having to involve his 'own' Immigration Department. To give this police department the job of registering details would be asking a lot. Moreover, given the tense atmosphere provoked by the incident with the six Ghanaians, any call to register with the Immigration Department would probably have been counter-productive. It was certainly discussed in the Co-ordination Centre that this department should be kept in the background as much as possible during activities to compile what soon came to be known as the Kosto list (initially known as the Wildekamp list).
16 Social Services and Housing Departments interviewees.
17 Housing Department interviewees; Nieuw Amsterdam Housing Cooperative interviewees.
18 A similar example of faulty communication between the city Housing Department and the Co-ordination Centre is noted in Chapter 3: the department dutifully kept faxing its lists of people who had been 'found', and could not understand why the centre's casualty figures remained so high. They did not question whether or to what extent these lists were noted by the Co-ordination Centre and the police, but took it for granted.
19 Social Services Department, Evaluatieverslag (Ramp)hulpverlening Vliegtuigongeluk in de Bijlmermeer (Evaluation report), 4.
20 Quoted from *De Volkskrant*, 7 November 1992.
21 *Het Parool*, 17 October 1992.
22 For details of the 'street-level bureaucracy' concept see: M. Lipsky (1980). *Street-Level Bureaucracy: Dilemmas of the Individual in Public Services*, Russell Sage Foundation, New York.
23 The Housing Department's first work meeting on 5 October had immediately drafted norms whereby staff could decide who was and who was not entitled to replacement housing. These were described as: 'workable norms ... justifiable norms, strict but also fair norms', Denijs, op. cit.. This business-like and formal approach was considered necessary to separate possible abusers from victims entitled to help. This was also prompted by reports from staff at the sports hall, within a few days, of 'free riders' among the applicants. Once initial emergency help had been organised, there was also an awareness at Social Services that not everyone who claimed to be a victim should be given the requested assistance, without some checks being made.
24 Extension of the registration deadline was also needed to meet the legal administrative principle of equal treatment for equal cases.
25 *Het NRC Handelsblad*, 5 April 1993.
26 Chief of Police Nordholt in *De Telegraaf*, 15 October 1992.

CHAPTER 6

Crisis decision making: the Co-ordination Centre

Introduction

This chapter describes and appraises the performance of the Co-ordination Centre. Our approach is to focus on the four crucial functions actually fulfilled by the Co-ordination Centre during the Bijlmer disaster:

- Estimating and defining the nature and seriousness of the disaster and appropriate action by the city authorities and services;
- Exerting general leadership of the crisis response during the period when the disaster declaration was in force, including the determination of relief strategy and longer-term activities;
- Managing and co-ordinating outgoing information for the general public and media;
- Symbolically representing and dealing with the collective stress caused by the disaster.

Each of these functions will be treated in more detail in separate paragraphs, with due attention to both stronger and weaker aspects of performance. This will be followed by a more detailed analysis of the Centre's decision-making process in terms of three dimensions regularly employed in similar Crisis Research Center studies, namely: organisation, information processing and psychological aspects.[1]

Official definition of the situation

Channelling crisis perceptions: basic notions

To establish a potent and co-ordinated approach to a disaster, it is very important for the authorities and services to work from a more-or-less shared view of the nature and seriousness of events. In the formal sense, the Mayor's disaster declaration and decision to have the Co-ordination Centre become operational is an important tool. This tells everyone involved that the situation is classified as serious and large-scale. Declaring the situation a disaster provides the Mayor with a number of far-reaching additional powers. The usual style of government by the Board of Mayor and Aldermen, which is often dominated by the latter and characterised by inter-departmental and party political differences, temporarily makes way for a centralised administration with the Mayor in a more hierarchical relationship to the Aldermen. The disaster declaration also allocates special tasks to the Chief Fire Officer.

The official definition and formal classification of the situation may alter during the course of a disaster. Obviously, disasters are not purely physical events; they are also situations of collective stress (see Introduction). A disaster has an impact on a community of individuals and groupings; the apparatus of government, together with private sector relief bodies, endeavours to support the affected community. But it is hard to predict accurately the nature of the various reactions from the community, and the specific relation between the community and the official response to the disaster. These factors can change continually. In turn this can mean a shift in the nature of issues as the disaster period runs its course. This requires that the authorities constantly update and complement their picture of the situation, in line with actual developments. Where required, they should then adjust the content or form of their response appropriately.

A second situation, where classification does not stop with the disaster declaration, is as obvious as it is sensitive. At a given moment the decision must be taken that, at least from the administration's angle, the disaster is sufficiently under control to withdraw the disaster declaration. The city's formal administrative organisation then returns to 'normal' – from strongly centralised authority to the collegial system. This decision – and more particularly its timing – may be sensitive from both political and social angles. On one hand, allowing the formal state of disaster to go on for a longer period imposes a heavy burden on the departments and operational services involved, and may encounter objections from the Aldermen and/or Councillors whose authorities and influence are curtailed under such a regime. On the other hand, sections of the community affected may take an early withdrawal of the disaster declaration as a sign of City Hall's indifference to the suffering caused by the disaster and the gravity of the resulting problems.

Classifying the situation: the disaster declaration

The crash of the El Al Boeing 747 on a residential complex, with the immediately visible results in the shape of an inferno and collapsed buildings, left the authorities little scope for a cautious approach and evaluation. With other types of major incidents, like fires and explosions at industrial sites, the appropriateness of a disaster declaration may not be so clearly or urgently apparent; it is contingent upon the actual location in relation to residential areas, the presence of a company fire brigade and often inadequate information from the company.

In this particular case the Mayor declared a disaster situation very soon after the aircraft crashed. This was reflected in the speed with which the Co-ordination Centre at City Hall was made operational. The speed of mobilisation was helped by the fact that a significant number of staff live in the immediate vicinity of City Hall. The crash may have happened at the weekend, but the fact that it was on Sunday night did mean that most people involved could be reached at home. Nevertheless, some were far away from Amsterdam. Even so, their absence was adequately compensated for.

The centralisation of authority on the Mayor caused no problems for the Board of Mayor and Aldermen. The Board showed a high degree of collegiality and team spirit, with the Mayor's leadership taken for granted. It proved possible to involve Aldermen *ad hoc* in the supervision of activities of departments that were formally part of the portfolio of fellow Aldermen.

The dynamic of the situation: a time lag

After the crash of the 747, initial attention in the Co-ordination Centre focused exclusively on putting out the fire, attempts at rescue, and evacuation and relief of the survivors. Due attention also went to maintenance of public order, including action to deal efficiently with the 'mass assault' on the disaster area by services, volunteers, media and the public (see Chapter 2). This was reflected in the organisation of the Co-ordination Centre: alongside the Mayor the key positions were taken by Fire Chief Ernst, Chief of Police Nordholt and the General Administrative and Legal Affairs Department (ABJZ).

In the Co-ordination Centre, during the first evening and night, the initial classification of the disaster situation continued to be in the direction of a major accident. During the first few days, attention was strongly concentrated on the problem of recovery and identification of human remains. It soon became clear that determining the precise number of victims would be highly problematic. Several interviewees would later describe the decision taken on Tuesday morning to accelerate recovery as the most important and difficult one taken in the Co-ordination Centre at that time. Considerable energy was also required to get a grip on the organisation of relief in and around the Bijlmer Sports Centre. The need to arrange a collective ceremony of mourning, and perhaps also a funeral later in the week, was taken into account from the very start.

From Wednesday 7 October onwards, all these considerations focused on a single point, namely the greatest possible care in preparing for X-Hour, the time set for release of the provisional lists of dead and missing. With this in mind, the Co-ordination Centre took several measures, such as obtaining an outside, second opinion and a critical examination of the results of the missing persons enquiry, before making them public (see Chapter 3).

Meanwhile, several city departments had been active since as early as Monday morning, not so much with countering the immediate consequences of the accident, as providing more structural relief to the victims. This was notably the case for the Housing Department and the drafting of the after-care plan under the responsibility of the Medical Services Department. Partly because the Housing Department was not present in the Co-ordination Centre, it proved problematic to get the administrative definition of the situation adjusted. Although initially the Co-ordination Centre concentrated strongly on the issue of the death toll (the narrower side of the disaster), the question of more structural help for a hard-to-define group of survivors started to come to the forefront (the local crisis caused by the disaster). Material relief of the survivors (somewhere to live, benefit payments, dealing with illegals) gradually developed into a considerable source of tension between relief workers and those seeking help. There were also co-ordination problems between the city departments and the Southeast Neighbourhood Council. The initial decision to make Social Services responsible for material relief, and the absence of the Housing Department from the Co-ordination Centre, eventually led to a situation whereby the Co-ordination Centre was not fully or adequately aware of the issue. This caused a time lag between the change in the main thrust of events in and around the Bijlmer (from a typical 'disaster' to a more complex urban 'crisis') and the authorities' awareness of these developments in the Co-ordination Centre.

End of the crisis: withdrawal of the disaster declaration

The Co-ordination Centre successfully avoided major friction around the withdrawal of the disaster declaration and the scale-down of the disaster organisation. When the dispute arose on closure of the sports hall, the decision from the Co-ordination Centre was that it should stay open longer. The Mayor did not hesitate to insist that departmental heads comply with this decision. More generally, there was a strong feeling in the Co-ordination Centre that premature 'routinisation' should be guarded against, given the threat of fragmentation this might pose for relief sources and loss of mutual co-ordination between the various bodies involved. This could mean the victims falling into the gap between emergency help and more structural forms of relief and care. And, for this reason, the disaster declaration stayed in operation longer than originally envisaged.

Allowing the disaster declaration to remain in force until 28 October did not create any noticeable difficulties as far as the City Council was concerned. More generally the actions of the Co-ordination Centre appear to have enjoyed full council support at all times. This support was encouraged by an early tour of the disaster site for council members; with their own eyes they absorbed the facts underlying the Co-ordination Centre's definition of the situation. This made any discussion of the gravity of the circumstances and the legitimacy of extraordinary governmental measures redundant.

The eventual decision to withdraw the disaster declaration appears to have been prompted by two incidental factors, rather than by any politically sensitive considerations: firstly the noticeable physical and psychological fatigue of several members of the Co-ordination Centre; and secondly the growing pressure on Co-ordination Centre members – partly from sections of the city apparatus not involved in the disaster – to resume their normal duties again. After more than a week (for some people more than two weeks) of extremely intensive labour, the operation had to be brought to an end. Nonetheless, actually achieving this was not easy. Several interviewees in and around the Co-ordination Centre noted a certain tendency to disaster addiction: people worked in a frenzy and it was hard to break free from it. It was hard to hand over responsibility for the care of the victims to the proper organisations. It meant refocusing attention on the mundane tasks of city management, living with the more complicated administrative relationships – in short, returning to the order of the day, which, it seemed, would never be the same again.

Overall direction and co-ordination

Crisis co-ordination: basic notions

An important function of a co-ordination centre is to shape governmental action in response to a disaster. The co-ordination centre is the place where crucial strategic decisions should be taken on the nature and design of the response. It should also be the location for governmental co-ordination of action by the various services involved in the disaster. At the same time, there needs to be a differentiation between the *internal* co-ordination among the various city services, departments and bodies, and the city's *external* co-ordination with other levels of government, private bodies and – in this case – authorities abroad.

Research into strategic decision making in crisis situations demonstrates that it is often very difficult for a central decision group, like a Co-ordination Centre, to get a grip on events on the spot. There is often a gap between governmental decisions and on-the-spot implementation. At the top, the authorities' frame of reference pertains to general strategy; they try to influence and set standards for the public and political perceptions of the crisis. At the level of implementation it is of paramount importance to have a keen eye on the far-reaching, often traumatising effects crises have on people, and the dilemmas faced when general policy measures have to be applied to real-life situations. Moreover, in highly dynamic crisis situations, there may well be occasions when the officials responsible for implementation – sometimes fighting against the clock – cannot avoid taking independent decisions which later prove to be of major significance. The outcome of both developments is that, at given moments, co-ordination centres may lose their overview and their control of events.[2]

The contrasting risk is that of 'micro-management'. Then, the authorities in a co-ordination centre become fixated on details of implementation and interfere in the territory of the operational leadership (the 'I-was-in-the-army' syndrome). This extends into the problem of strategic co-ordination. Large-scale incidents tend to put heavy pressure on the co-ordination function. A famous US crisis researcher goes as far as concluding that, in many crisis situations, co-ordination is the problem rather than the solution.[3] Some departments and services feel uncomfortable in the type of integral project organisations that materialise in crisis situations, and go their own way. Sometimes there is conflict on the operational level. Services unfamiliar with the others' background, expertise and procedures must suddenly and closely attune their actions. That causes friction, as expressed in the Dutch saying 'unknown is undervalued'. During a crisis, old bureaupolitical tensions may arise and lead to vicious fights. Some organisations may use crises to score points, maybe at the cost of others. Perhaps cynically, crisis experts call this the 'battle of the Samaritans'. Interventions from other levels of authority may further complicate matters.

Leadership by the Co-ordination Centre: strategic decisions

With a few exceptions (see Chapter 5), the general picture of the disaster period is that most of the important decisions were dealt with in the Co-ordination Centre, and taken by its members. The key decisions in this category were the:

- Strategic decision by the Mayor to declare the situation a disaster, immediately after he had been informed of the accident;
- Decision to treat all victims always equally, irrespective of whether or not they were illegal immigrants;
- Decision to hold twice-daily press conferences, preceded by central briefings in the Co-ordination Centre;
- Decision to have the Social Services Department co-ordinate material relief to victims;
- Decision to accelerate the recovery of human remains;
- Fixing of the date and time at which the list of victims would be released;
- Decision not to give interim reports on the number of missing persons, prior to X-Hour;
- Decision to invite the British experts;

- Decision on the drafting and main elements of the after-care plan;
- Decision to have the Southeast Neighbourhood Council co-ordinate in the Bijlmer Sports Centre;
- Various decisions on the duration and form of relief in the Bijlmer Sports Centre;
- Decision to move large numbers of victims from the Bijlmer Sports Centre and overnight centres to the Novotel;
- Decision not to have the police intervene at the Park Lake Hotel, following reports of aggression towards Social Services officials;
- Decision to withdraw the disaster declaration.

Most of these decisions were eventually taken by the Mayor, usually after extensive consultation in the Co-ordination Centre. Two important items are missing from this list of strategic decisions: the decision to regularise those illegal immigrant victims who met certain conditions, and the decision to extend the registration period for illegal immigrant victims. Given the far-reaching implications of these steps, more detailed discussion with all parties in the Co-ordination Centre would have been beneficial. Had such discussion taken place, more attention would probably have been paid to unintended but potentially negative ramifications (the rush of people to the offices of the City Population Register) and technical complications of implementation (delicate procedures on eligibility).

The decision to prolong the deadline for registration for the Kosto list was a remarkable example of an operational decision – taken under extreme pressure in terms of time and workload – producing unforeseen strategic consequences. Despite the understandable reasons for the decision and the time pressure involved, consultation in the Co-ordination Centre was obviously needed. Had it occurred, it might well have been decided to complete registration by forcing up the tempo of work. Then again, the Co-ordination Centre might not have realised the full implications of shifting the deadlines, and itself have reached the same conclusions. Even so, the decision would still have been taken in the appropriate place – rather than coming as a shock to members of the Co-ordination Centre during the course of the following week.

Internal co-ordination

The Co-ordination Centre operated along strictly controlled lines. The rhythm of a central briefing prior to each conference – which came about more or less by chance – ensured a constant, overall update for those in authority. This strict rhythm and direction of activities in the Co-ordination Centre generally ensured better communications between the authorities and the implementing level than in most large-scale crisis situations. Regular inspection trips to the crash site by the Mayor and police and fire chiefs also contributed and kept them in touch with reality.

Despite the strict management of briefings in City Hall, the actual co-ordination of action by most of the services and departments involved took place largely outside the Co-ordination Centre, unhampered by any sudden interference from City Hall.

For instance, the Disasters Act put the fire brigade in charge of co-ordination of emergency services at the site. Initially, during the first few days, this aspect was the cause of some friction between the fire brigade, city police and the RIT. This was caused by the very limited space in which they had to work, their varying methods and lack of familiarity with each other's specialisms. However, following the decision to accelerate recovery and streamline the identification procedures, co-ordination on the spot proceeded without any notable problems. In any case, the deadline set by the Co-ordination Centre for completion of the recovery work was met with plenty of time to spare.

In another example, the Co-ordination Centre had allocated the actual co-ordination of material relief to the Social Services Department in an early stage. Some aspects of the co-ordination were sub-optimal and caused a series of mutual communication breakdowns. Also, at times, the Housing Department or the Social Services got the impression that the Co-ordination Centre had insufficient insights into practical implementation. This was exacerbated by the fact that the Director of the Housing Department was not a member of the Co-ordination Centre. This was corrected at a later stage.

When the situation in the Bijlmer Sports Centre appeared to be getting out of hand, co-ordination of relief was allocated to the Southeast Neighbourhood Council. This decision was partly designed to increase clarity in the relationship between the Co-ordination Centre and the Neighbourhood Council – which had been less than perfect during the first few days of the disaster. Initially, the Southeast Neighbourhood Council did not have a strong enough representation in the Co-ordination Centre. During this period there was no clear demarcation of powers and responsibilities.

In practice, however, the Co-ordination Centre assumed control. Basically, the Co-ordination Centre decided and the Southeast Neighbourhood Council implemented these decisions, while carrying out a range of activities in and around the Bijlmer Sports Centre.

Allocation of the co-ordination task for the Sports Centre also needs to be viewed in this context. Following this intervention, the Southeast Neighbourhood Council appeared largely to go along with the procedure that had arisen. However, the Southeast Neighbourhood Council did make a definite stand on the dispute about timing of the run-down of activities in the sports hall, and eventually found that the Co-ordination Centre came over to its point of view.

Compared to major disasters elsewhere, the effectiveness of the internal co-ordination of Amsterdam's authorities and services was overwhelmingly positive. Problems of co-ordination are inevitable where a large number of operational departments and services are concentrated in a very small territory. The need for stringent and accurate co-ordination of usually very different, separate activities and constant heavy time pressure create difficult conditions.

The Co-ordination Centre gradually developed an intricate 'co-ordination mix' to cope with the situation. This comprised: division and delegation of co-ordination tasks to given bodies; stringent monitoring of progress via the briefings; unambiguous intervention in the event of friction or deadlock.

From a governmental perspective this co-ordination style worked well. In general the departments involved and the Co-ordination Centre achieved reasonable operational co-ordination. Many interviewees classified the mutual relationships in terms of consensus, rather than conflict – let alone 'bureaucratic warfare'. Inter-organisational misunderstandings and frictions did show up here and there; but these caused no real harm to the effectiveness and efficiency of services provided by the city. It is certainly true that internal fine-tuning did involve difficult meetings, irritations, some raised voices, and one formal order to comply with a previously declined instruction. But compared with other crisis situations, this was a small price to pay.

External co-ordination

The external co-ordination picture is more ambiguous. It is favourable as far as relations between City Hall and other levels of government are concerned. Firstly, the role of central and provincial government was limited to providing support and a number of good services; the major part of the counter-disaster operation was led from Amsterdam. Supra-local activities could be kept to a minimum precisely because the disaster happened in a major capital city – Amsterdam – which has at its disposal large-scale, professional and specialist resources. A comparable disaster in a smaller municipality would have demanded a far greater role for other levels of government, and far more comprehensive outside assistance.

Secondly, contacts with the Province of North Holland, and the Interior Ministry's National Co-ordination Centre in particular, were relatively smooth. The Amsterdam Co-ordination Centre kept these two parties constantly updated on developments, whether or not news was actually sought, and provided assistance where required. The same applied to the Prime Minister's staff. The Ministry of Transport and Public Works – in particular the Civil Aviation Authority – also gave full co-operation. Nonetheless, it still demanded considerable effort on the part of the Co-ordination Centre to clear the airspace above the Bijlmer, after the disaster, and to keep it clear. Contacts with diplomats representing most of the countries involved (e.g. Israel, Surinam and Ghana) went well, helped by the mediation of the National Co-ordination Centre in The Hague.

And City Hall's insurance section smoothly established contacts with the various private insurers at home and abroad.

All this contrasted with a number of problems at the level of external co-ordination. The 'elite assault' on the disaster area was no less of a nuisance for being comparatively innocent: authorities, experts and other official-looking parties from home and abroad queued up to visit the accident site and the Co-ordination Centre. The Co-ordination Centre log book shows that arranging such visits took up much time, particularly during the first few days of the disaster period. Naturally, interest from the Dutch Royal Family and the Cabinet was greatly appreciated and provided considerable support for all involved. The decision to show the authorities from The Hague around the site in one single 'batch' implied complex scheduling and logistics. Subsequently, problems arose when yet more officials from The Hague and elsewhere began to apply. At a certain moment the decision was taken in the Co-ordination Centre to halt such visits. A number of would-be visitors felt they had been treated with less than due regard, and complained.

Contacts with representatives of the government of the Netherlands Antilles were downright difficult. The Antilleans were very concerned about the fate of the large number of their people living in the Bijlmer. Apparently, the Antillean authorities also lacked confidence in the way City Hall and the Amsterdam services were dealing with the issue. During her visit, the Antillean Prime Minister, Mrs Liberia-Peters, was critical of the way the Co-ordination Centre was dealing with publication of the list of victims. The Antilleans then made their own list, which was submitted to the Co-ordination Centre and included in the police investigation. At a certain moment, there was even a suggestion from the Antillean side that this list might be made public if the Co-ordination Centre failed to release its own lists. In the meantime, an Amsterdam police officer was sent to the Antilles to consult with his opposite numbers on the spot, and to attempt to verify their list. This visit was marked by good co-operation between fellow policemen (see Chapter 3).

Another point of tension between the Amsterdam and Antillean authorities concerned the relief and assistance given to victims. At a certain moment, the Antillean government, at its own initiative and expense, offered victims from its community and their families accommodation in the Park Lake Hotel. FORSA, the Antillean interests association in The Netherlands, was given the task of co-ordinating relief at this hotel. This created, as it were, a second relief circuit, operating beyond, and sometimes against, City Hall's policies. Later, the Antilleans unilaterally ceased paying for the Park Lake Hotel. All this did nothing to promote an optimal joint effort for the benefit of the victims, whether Antillean or not.

Information and media management

Public information during crises: basic notions

In a large-scale crisis it is hard to underestimate the significance of the mass media for the course and outcome of the event, as well as the formation of post-event perceptions. Crises on the scale of the Bijlmer disaster immediately provoke a determined mass invasion by the domestic and foreign media. Their reporting of events is crucially important in dissiminating ideas of what actually happened, among the community at large, abroad, and often for a large number of local and national politicians. It also influences the assessments of the authorities' performance. Moreover, the mass media can fulfil an essential role in informing the public. They can both promote and effectively help quash rumours.

All this makes information and media management an essential component of crisis management. In other words, the operational arena is not the only place where the success or failure of crisis management is determined. The way the authorities manage their flows of outgoing information also counts – to an even greater degree. In preparing for and performing during crises, many policy-makers and administrators appear to give insufficient priority to this possibly unwelcome but nonetheless essential aspect. They view the mass media exclusively in terms of a nuisance, see press conferences as a necessary evil and devote all their energies and attention to the substance of measures to be taken, meanwhile neglecting their presentation. They take insufficient account of the needs and demands of the mass media. In turn this deprives them of an important means of gaining control over events while giving a boost to rumour-mongering and negative perceptions. Obviously, if the press are not properly provided with information, they will go out looking for it themselves, and in a large-scale crisis situation there are always plenty of people ready to talk about the whole spectrum of events from a wide variety of angles and personal views. An active, professional and fair media relations programme can often make a valuable contribution to meaningful crisis management.[4]

The triumvirate in action

It was clear that the pivotal task of co-ordinating information and media contacts was a responsibility of the Co-ordination Centre, backed up by the City's Information and Public Relations Department. The triumvirate of Mayor, Chief of Police and Chief Fire Officer acted as the permanent public face of the Co-ordination Centre and hence the City of Amsterdam. Its performance can be characterised as follows.

Firstly, on the first evening, with one exception, City Hall spoke with a single voice, that of the Mayor, backed up by the Chiefs of Police and the Fire

Brigade. The authorities soon learned the lesson of the first evening when the media were also obtaining news from a large number of *ad hoc*, semi- or indeed unofficial spokespeople from a variety of organisations, often providing contradictory, dubious or downright inaccurate information (e.g. the announcement that 12 bodies had already been recovered). From then on, close co-operation was established between most of the information/press relations departments involved.

In addition, there was a definite rhythm. After the improvised and somewhat chaotic initial press conference around midnight on 4 October, a regular schedule was rapidly established, with at least two press conferences daily, at set times in the morning and early evening. As previously noted, this rhythm seems also to have had a particularly beneficial effect on the workings of the Co-ordination Centre and indeed the broader counter-disaster operation. In relation to the mass media, this created a predictable and controllable time and place pattern. It gave the press points of reference with intervals to seek news in the field. The rhythm and timings seem to have suited media deadlines.

Furthermore, the authorities said what they had to say clearly, telling neither more nor less than they were inclined. In the inevitable confusion of uncertainty and burgeoning rumours around the number of dead and missing, the Co-ordination Centre mapped out a course and followed it systematically (announcement of the results of recovery, identification and the police missing persons enquiries, at X-Hour set by the Co-ordination Centre). However, it was certainly a nuisance that information on the list leaked well before this deadline, on Friday 9 October.

Moreover, a good balance was struck between emotions and sentiments on the one hand, and facts, figures and procedures on the other. The *leitmotif* of 'caring government' emanated from the circles around the Mayor: government doing its best to soften the blow and help people in distress. Judging by the public opinion poll's positive evaluation of the city authorities' performance (including information management) during the Bijlmer disaster, they successfully conveyed this impression to the citizens of Amsterdam in word and deed (see Appendix 2).

Finally, press conferences were adequately utilised to manage incidents. There were some misunderstandings over the implementation of decisions taken by the Co-ordination Centre. Left uncorrected, these would have provoked negative rumours and unnecessary frustration and anxiety among the victims and their families. Two episodes stand out in particular: the Social Services' use of standard forms with pay-back clauses, and the registration of the six illegal Ghanaians by the Police Immigration Department, despite assurances from the Co-ordination Centre that relief would be provided to everyone irrespective of residence status. In both cases, not only were measures taken to prevent repetition, but the Co-ordination Centre also dealt with the issues openly and at length during the press conferences. In addition, every effort was made to counter inaccurate rumours and false conceptions at an early stage. At the same time, it is hard to judge how far this was successful. The two incidents provoked a shock-wave across the illegal immigrant groupings and their relief organisations – who were already, by virtue of their circumstances, distrustful of the authorities.

Some problems

This generally positive picture does not mean that there were no problems with regard to information activities. A number of complications deserve a mention. First of all, information activities on the first evening were unco-ordinated. The Southeast Neighbourhood Council held a press conference on its own initiative, and representatives of various departments answered press enquiries without any mutual agreements on or verification of the information given. This was partly due to the incorrect assumption that one had to have something to announce be fore calling a central press conference.[5] This ignored the fact that, when a disaster is at its freshest, everything is newsworthy. Moreover, the speedy establishment of a centralised system and fixed rhythm in briefing the press helps to rectify nascent rumours.

Initially also, the international dimension was not given the attention it deserved. During the second press conference, the foreign press were clearly annoyed that proceedings were in Dutch and that there were no interpreters. This omission was later put right.

At and around the disaster site, and later also at the Bijlmer Sports Centre, there were some unpleasant clashes between journalists and members of the emergency services, notably police and fire-fighters. The press felt they were being kept at too great a distance, and the people in uniform thought the press were being far too pushy, and that they were getting in the way. Later, organised tours of the disaster site by the various services' press officers did much to counter this friction.

The strict regime of the press conferences was only broken on one occasion, when State Secretary of Justice Kosto called a press conference – following his meeting with Amsterdam Police Chief Nordholt – to announce the decision on a regularisation procedure for illegal immigrants among the victims. This took some people in Amsterdam by surprise.

Management of collective stress

Channelling emotions: basic notions

Public information and press relations are an integral and important part of crisis management. The same applies to coping with what might be called the social, emotional and symbolic aspects of disasters. The importance of sensitivity by the authorities in this area relates directly to the concept of a disaster as a collective stress situation. Parts of the community are disrupted as are the routine patterns of the overall community and the political order. As well as demanding concrete and extraordinary measures, this also requires an administrative style and presentation clearly demonstrating that the authorities are deeply conscious of the enormity of events and their social ramifications. Disaster and other large-scale crisis situations also demand a form of socio-emotional leadership: a government that is aware of the feelings of the community and acts appropriately. This dimension of crisis response is hard to translate into disaster planning. The nature and gravity of the disaster-related collective stress varies from case to case. Adequate socio-emotional leadership demands that the authorities keep their finger on the community's pulse. This is one of the most visible components of the social and political burden carried by the authorities.

Crisis research demonstrates authorities' frequent difficulty in dealing with the socio-emotional dimension of crisis management.[6] Some adopt a no-nonsense style and are then later surprised at accusations of heartlessness. Others find it hard to strike the right balance between a show of respect for a display of collective emotions, and the need to take unconventional, occasionally downright tough, damage-limitation measures.

The Amsterdam experience with 'caring government'

In several respects the composition of the Co-ordination Centre reflected the diverse aspects of crisis management. There was a mix of people who were by definition alert to social and political currents. The Mayor in particular emphasised empathy and the psychological aspects of the exercise of authority. It is significant that when more restrictive, 'business-like' and more generous, 'empathetic' approaches were in danger of clashing – as was the case during decision making on the scaling down of facilities in the Sports Centre, or in the discussion of strict versus pragmatic criteria for the status of the victims – the Mayor always gave priority to empathy. Naturally, it was hardly surprising that at certain moments this attitude caused friction in the Co-ordination Centre. For that matter, in view of the Mayor's position in the Co-ordination Centre, it was hardly astonishing that the outcome of these disputes also favoured the psycho-social line.

During the course of the disaster period, some people in the Co-ordination Centre started to adopt the 'caring government' concept borrowed from the Lockerbie experts. This meant that City Hall's response to the disaster would clearly demonstrate the best possible efforts to provide material and psychological relief for those affected, in a fair and, above all, committed manner. The Co-ordination Centre showed its commitment to the concept of caring government in a variety of ways:

- Ongoing efforts not to discriminate on grounds of immigration status when providing material and intangible relief (some city departments, like Housing, did not need guidance on this; others, like the police, did);
- Encouragement of the strictest possible link between material and intangible relief (keeping a broad service package in position, support for psycho-medical arguments on the Sports Centre's symbolic function);
- The policy of having at least one Alderman to represent the city at every funeral of a disaster victim, and active encouragement of involvement by Councillors in this regard;[7]
- The major effort from City Hall – in consultation with the groupings representing the victims – devoted to the organisation of the collective service of remembrance and a possible collective funeral service (which later proved unnecessary);
- The frequent visits by the triumvirate of Mayor, Police Chief and Fire Chief to the disaster site, and their visible empathy with the teams performing the onerous and difficult task of recovering and identifying human remains;
- The relatively high priority put on expressing appreciation to, and providing after-care for, the personnel of the various services involved.

While there can be no doubt of City Hall's success in demonstrating its willingness and ability to be a caring government, it is a philosophy that needs to be approached with caution. The Bijlmer disaster also showed that – caring or not – clear boundaries need to be defined for government care. Indeed, many of the problems and much of the negative opinion that began to arise around the victims of the Bijlmer disaster were rooted in the vague or contradictory criteria formulated by the authorities for the status of victim; unintentionally, but also inevitably and predictably, this provoked 'free-riding' behaviour by some pseudo-victims.

In addition, the overall concept of caring government embraces many different ways of providing that help. Merely emphasising the care aspect will not automatically banish the differing approaches to their duties among the various city services. There were differing insights from department to department when it came to the provision of psycho-social and material care. From the very start of the Bijlmer disaster, organisations like the Housing Department, and to a lesser extent the Social Services Department – which both had the job of trying to allocate scarce government resources on a day-to-day basis – took a far more formal approach, with more restrictive methods, than was realised in the Co-ordination Centre.

Besides, implementing 'caring government' assumes that the local co-ordination centre has the means independently to realise the philosophy. In a city the size of Amsterdam, with all its resources in place, it is a reasonable proposition. But in many other places in the Netherlands the local co-ordination centre would depend strongly on the willingness and ability of other levels of government to provide the required broad package of goods and services. This can soon delay or otherwise complicate their provision. With the Bijlmer disaster, there was also the issue of immigration policy, which was formally outside the authority of the Mayor, and which – after discussion within a very small group – led to the Chief of Police approaching the State Secretary of Justice. The nature and scale of the disaster impose limits on what can reasonably be expected from – and thus offered by – caring local government. To a degree, this aspect also makes the Bijlmer an atypical disaster: the municipality involved was largely able to handle the situation autonomously. This is an exception rather than a rule.

Finally, the philosophy of 'caring government' imposes a heavy burden on the authorities. They are expected constantly, in word and deed, to empathise with the widespread human suffering caused by a disaster like that in the Bijlmer. At the same time they are expected to act coolly, resolutely and, where called for, with a degree of toughness. This combination of tasks calls for an uncommon measure of emotional resilience – something that becomes increasingly difficult the longer the disaster drags on.

Co-ordination Centre in action: collective choice under pressure

Organisation and modus operandi

The Co-ordination Centre acted as a forum from which the Mayor and others in authority gave guidance on the management of the disaster. In addition – and this had not been foreseen in contingency planning – some of the agency heads and other officials used it as a base from which to direct their organisations. It also became the focal point for communication from various units in these organisations. This applied in particular to the police and fire brigade.

One of the outcomes was that a considerable number of officials became part of the Co-ordination Centre, or spent time there on a regular basis during the period of the disaster. About 60 people were involved. While at first sight that appears an unmanageably large number, it has to be borne in mind that the intensity of events demanded that for the first few days the Co-ordination Centre had to be fully staffed around the clock; this was later reduced. Secondly, within this overall 'membership', there was a very small core team of the most senior executives with prime responsibilities. They were the ones most intensively involved and were surrounded by a much larger and changing group of officials.

1 *Logistics*. The physical infrastructure of the Co-ordination Centre appears to have served the decision-makers well. Even so, several aspects deserve to be noted:

Access. Co-ordination centres exert a definite pull on the authorities and officials. This was also true of the Bijlmer disaster. Certainly in the initial hours and days, there was a major influx of people. For example, the police often had more than five people present, mainly because a significant part of the police operation was led by Chief of Police Nordholt and other senior officers from within the Co-ordination Centre. That soon added to the numbers present, and this could have been a hindrance in case of considerable time pressure. However, truly high-speed emergency decision-making was virtually absent during the Bijlmer disaster, with the exception of the incidents of aggression towards Social Services officials in the Park Lake Hotel. It could be a different story in a future crisis situation; large numbers of people in the Co-ordination Centre could well create a more direct nuisance.

Prominent positioning of TV monitors. Television is an extraordinarily powerful medium. Particularly during crisis situations, decision-makers can easily get the impression that the dramatic images on the screen actually represent the most important reality of the crisis. In fact, TV pictures generally present a very limited, even distorted, picture. Decision-makers in co-ordination centres who indiscriminately accept TV pictures as the basis for their information, take decisions founded on an inaccurate impression.[8] This calls into question the location and use of TV monitors in the Co-ordination Centre, showing details of recovery work during the first few days. This tended to encourage micro-management of the recovery operation, to the surprise of people in charge on the spot. Moreover, the detailed shots of human remains and other harrowing scenes appear to have had a stress-enhancing effect on some people in the Co-ordination Centre, without actually providing essential information. Lastly, the fact that the decision-makers in the Co-ordination Centre spent more time than was strictly necessary on events at and around the disaster site, during the first few days, may be partly attributable to the impact of the pictures on the TV monitors.

2 *Composition.* Aside from the numbers present, the stature of people on duty at the Co-ordination Centre also deserves further consideration. The two most striking problems in this regard are the tardy and initially lightweight presence of the Southeast Neighbourhood Council, and the absence of the Housing Department. The former certainly calls for a permanent solution, since Neighbourhood Councils are very likely to be involved in future disaster situations, in addition to the city in its entirety. In the case of the Housing Department, the usefulness of its presence in the Co-ordination Centre would depend more on the nature of a given disaster.

A more difficult question bears on the relative importance of the presence of departments and services. The physical configuration of the centre (two half-circles of tables, one behind the other, facing a central point where the monitors, maps and 'whiteboards' are positioned) would appear to suggest a system of first and second ranking for services. The 'first circle' of organisations (the police, fire brigade, and the General Administrative and Legal

Affairs Department or ABJZ) tends to be focused on issues of public order and safety. The 'second circle' of organisations (the Social Services, Medical Services, the City Population Register and the Housing Department) stands for another set of disciplines and orientations; these are far less incident- and much more policy-focused. We cannot avoid the impression that during the first week this second circle was far less in the picture than might have been desirable (e.g. with an eye to a faster grip on the missing persons issue, and the timely signalling of the development from disaster to crisis). The situation became more balanced later on, a good example being the fire brigade's moving into a less central position, at its own initiative.

3 *Leadership.* Right from the start the Mayor successfully combined and exercised differing leadership roles. He put himself at the centre of substantive decision making, and – through the General Administrative and Legal Affairs Department – imposed a stringent working pattern (in particular around the briefings); this could be designated as task-oriented leadership. He also attempted to regulate the atmosphere in the team, and to engineer dissipation of tension at crucial moments: social-emotional leadership. This combination of qualities made him an excellent but hard to replace leader. The vulnerability of this arrangement was demonstrated in the Mayor's absence: it was not always easy for his deputy to lead with the same authority, particularly when dealing with the outside world.

4 *Modus operandi.* The stringent scheme of plenary briefings had not been foreseen in contingency planning, but proved so successful that the idea deserves to be copied in the future. However, it was significant here that ABJZ'S management responsibility in this area meant a double task for the Head of its Public Order and Safety Unit, Sarucco. In the absence of the departmental director in the United States, she was, in addition to 'running' the briefings, also expected to give more substantive advice to the Mayor. The combination of tasks was a heavy burden for one person and could have led to conflicts of interest.

Information and communication

1 *Communications infrastructure.* Although the Co-ordination Centre is well provided with modern communications equipment – as in any disaster situation – there were great difficulties in establishing good communications with the disaster site during the first few hours. The inevitable overloading of the public telephone lines and the abrupt surge in demand as people tried to communicate with their organisations, created major problems.

2 *Rumours*. Partly due to the lack of adequate communications with the field, during the first evening and night the Co-ordination Centre was unable properly to sift fact from rumour. Subsequently, active measures were taken to verify rumours and rectify misunderstandings that could provoke rumour-mongering (repayment clause in the Social Services forms, registration of illegal immigrants by the police). In some cases it also transpired that there was a gap between policy aims and actual effects: for example, partly due to burgeoning rumours in the legal and illegal immigrant communities, these groupings were only partially convinced of the goodwill underlying the impassioned undertakings and pleas from the Mayor and police chief.

3 *Outside experts*. The discipline of the briefings helped the Mayor, in particular, actively and critically to evaluate incoming information. At several points the Co-ordination Centre sought to reduce in a responsible manner the great uncertainty surrounding crucial disaster-related issues. Notably, outside experts who had dealt with the Lockerbie disaster were called in to give a second opinion, and advice was sought from a trauma specialist. At the same time, obtaining information along this route was a sensitive issue *vis-à-vis* some of the relevant emergency services ('.... we know all this ourselves and can do it just as well....'). Although hurt feelings did not lead to demotivation or conflicts here, more extensive, drawn-out or publicly visible use of outside expertise might well have done so. The Lockerbie experts also appear to have made a psychological contribution, alongside their substantive input. Strangely, the actual presence of people who had been through similar events seems to have boosted morale. Moreover, Professor Busuttil's theory that some victims' bodies may well have been totally obliterated by the fire also offered a welcome – albeit in retrospect probably incorrect – explanation for what was at the time an unfathomable and disturbing riddle.

4 *Staffing*. There were considerable differences in the type and quality of information provision by the various city departments and services. Looked at from the other angle, information and instructions from the Co-ordination Centre reached the operational levels of the relevant organisations in a variety of ways. Obviously, internal staff's lines of communication were short and hence problem-free. The police and fire brigade were always able to ensure adequate two-way information. The police in particular had plenty of their people stationed in the Co-ordination Centre for this very purpose, and invested much time and effort in the process of translating policy to practice. Several other city departments, notably Social Services, appeared to lack this capability. On occasion, the lines between the Co-ordination Centre and the implementation level became blocked (i.e. the initial co-ordination decision did not get through properly to the rest of the organisation); and at other times these lines were far too short (i.e. the panic around the Park Lake Hotel).

Workload and psychological stress

In crisis situations, due to the pressures of work, decision-makers may experience stress which, in turn, may influence the quality of their thinking and actions.[9] It is far from easy to make robust statements on a phenomenon as complex as psychological stress. Hence, we can only give general indications, and these too should be treated with a degree of caution:

1 *Working hours.* During the Bijlmer disaster, most members of the Co-ordination Centre worked far too many hours in a day. While most officials positioned there were obviously used to long hours, in disaster situations like this the effect of the time factor on decision makers is exacerbated by the combination of unbroken and highly intensive activity under psychologically trying circumstances. There was no definite evidence of disturbing symptoms of over-tiredness or stress among the relevant decision-makers during the Bijlmer disaster. At the same time there appear to have been pronounced differences in the way those in authority coped with these situations. Several interviewees said that, after a while, notably after the disaster period had ended, they experienced more than their usual level of physical and mental fatigue. It is questionable, then, whether the rota system used would prevent a number of central decision-makers becoming exhausted during a more drawn-out crisis.

2 *The Mayor: role accumulation and media pressure.* In the Netherlands, the Mayor in particular is subject to extraordinarily heavy pressure in a crisis situation. On the one hand, he or she has a central role in the local governmental decision-making process. On the other hand, the Mayor is also the public face of the town or city, and as such must regularly confront the media. This means performing in a climate where rumour, recrimination and a whole variety of complications set the tone. Some Mayors are unequipped to handle this pressure. In this case, the performance of Amsterdam's Mayor was generally acclaimed for the balance between professionalism and empathy that he showed both on and off camera. But that balance is always precarious, and can soon come under pressure in a lengthier crisis, bringing more conflicts and more tragic choices. Given Amsterdam's present, strongly personalised crisis regime, these factors put the emotional side of the Mayor's personality to the test.

3 *Disaster addiction.* Many interviewees mentioned the difficulty they had in switching back to the old daily routine. This matches a well-established fact in the psychology of crisis decision making: the extraordinary circumstance of a crisis can put the people in authority in a sort of 'high', which is hard to shake off. Administrative lines are shortened. Bureaucratic culture is tempo-

rarily abandoned in favour of a pro-active approach. The decisions one takes have a directly observable result. In the case of the Bijlmer disaster this was reinforced by the authorities' strong sense of solidarity. There was deep sympathy with the suffering in the Bijlmer, and this made many executives in the Co-ordination Centre feel personally responsible. Because of this, there was a temporary moratorium on inter-departmental politicking. As the disaster period progressed, a number of developments started to erode communal solidarity with the people of the Bijlmer, and inter-organisational frictions rose apace. Even so, every effort was made in the Co-ordination Centre to maintain closed ranks as far as possible – without this actually obscuring the differences of opinion and the discrepancies. And so, when the disaster declaration was withdrawn, this made it extra difficult for some members of the Co-ordination Centre to return to the unyielding 'piecemeal' routines of a big city administration.

4 *After-care.* There does not appear to have been systematic after-care at the level of the Co-ordination Centre. And although it is questionable whether, and to what degree, this was necessary, several interviewees still had distinct feelings of unease. This is hardly surprising. The disaster demanded extraordinary effort, resolution, flexibility and tact from the members of the Co-ordination Centre. The strict regime also required concessions on the part of those in authority and heads of department. But perhaps the most frustrating thing was the realisation that, despite the overwhelmingly competent management of this disaster, hardly one single life could be saved. Neither were the authorities able to turn around the original, negative, stereotyped public perceptions of the Bijlmer and its residents, which returned following the initial period of widespread solidarity.

Notes

1 See also U. Rosenthal, P. 't Hart and M.T. Charles (1989). The world of crises and crisis management, in U. Rosenthal, M.T. Charles and P. 't Hart (Eds), *Coping with Crises: The Management of Disasters, Riots, and Terrorism*, 1–22, Charles C. Thomas, Springfield.

2 See P. 't Hart, U. Rosenthal and A. Kouzmin (1993). Crisis decision making: the centralization thesis revisited, *Administration and Society*, Vol. 25, No.1, 12–45, May.

3 E.L. Quarantelli (1988), Disaster crisis management: a summary of research findings, *Journal of Management Studies*, Vol. 25, 373–87.

4 Also see M. Kelly (1989). The seizure of the Turkish embassy in Ottawa: managing terrorism and the media, in U. Rosenthal, M.T. Charles and P. 't Hart (Eds), *Coping with Crises: The Management of Disasters, Riots and Terrorism*, 117–38, Charles C. Thomas, Springfield.; L.M. Wilkins, L. Wilkins and T. Walters (Eds) (1989). *Bad Tidings: Communication and Catastrophe*, Lawrence Erlbaum, Hillsdale.

5 Press, Information and Public Relations Department interviewee.

6 An object lesson is that of the Belgian Interior Minister, Nothomb, who was interviewed in his

Brussels apartment shortly after millions of TV viewers had watched the Heizel Stadium disaster unfold, and in a cool, almost analytical tone explained that he felt no need to go there himself. This, among other factors, made him the central target of the subsequent parliamentary enquiry. See P. 't Hart and B. Pijnenburg (1988). *Het Heizeldrama: Rampzalig Organiseren en Kritieke Beslissingen*, Samsom, Alphen. A less extreme example was the 1992 earthquake in Roermond, in the south of the Netherlands, when Interior Minister Mrs Dales' failure to visit the scene caused a local uproar.

7 Note that forecasts of the death toll were still high when this decision was taken. Downward adjustment to the much lower level would only come later.

8 This was a basic problem for the Co-ordination Centre at City Hall and for the police public order staff during the riots around Coronation Day on 30 April 1980. See U. Rosenthal (1989). A compulsive crisis: The inauguration of Queen Beatrix, in U. Rosenthal, M.T. Charles and P. 't Hart (Eds), *Coping with Crises: The Management of Disasters, Riots and Terrorism*, 225–51, Charles C. Thomas, Springfield.

9 See for example M. Brecher (1980). *Decisions in Crisis: Israel's Choices, 1967, 1973*, Transaction Books, New Brunswick.

CHAPTER 7

Urban crisis management in Amsterdam: conclusions

Several lessons can be learned from this analysis of counter-disaster and crisis management in Amsterdam following the aircrash in one of its most problem-ridden neighbourhoods. At one level, one may look back at the details of the operations and the policy decisions made, single out their strong and not so strong points, and develop specific recommendations for the Amsterdam authorities. At another level, however, one may take this particular disaster as an exemplar of a broader category of contemporary urban calamities, and reflect upon some of its generic themes and possible lessons for urban authorities and counter-disaster agencies in general. In this chapter, we shall take the latter approach and discuss the more general implications of the disaster episode. Yet, in order to do so properly, we shall at times provide some more detailed evaluations of certain features of the current case to illustrate the broader point.

At the same time, it should be remembered that several aspects of the Bijlmer disaster made it atypical, and therefore not in all respects a reliable basis for generalisations about urban counter-disaster and crisis management. For example, what is striking is the limited share of emergency relief and actual counter-disaster activities, contrasting with the 'blotting-paper' effect of the social problems exposed by the disaster. In this case, this was highlighted by the fact that the disaster hit an area with a disproportionately high immigrant population and raised the exceptionally complex issue of victims who were also illegal immigrants. With these caveats in mind, we draw five main conclusions, and present a number of additional discussion points.

Scale and capabilities

Conclusion 1: A metropolitan government should be able to handle most large-scale disasters on its own.

Modern cities are vulnerable to a wide variety of disruptions. The concentration of large numbers of people, the density of economic activities, the infrastructure of government, communications and transportation facilities, and the attraction of the big city for recreational purposes suggest ample possibilities for all kinds of threats and dangers to occur. A mass population living together in a confined space and exhibiting social, economic and political differences would appear to constitute a perfect setting for crises emanating from social and political tensions. This is part of the reason why urban history tends to be marked by social and political turbulence, riots and turmoil, if not coups d'état and revolutions. Yet modern cities are also vulnerable to exogenous disruptions that thrust themselves into the social and economic fabric of urban society and, on another level, damage shared expectations of the urban population at large about urban life as usual. Cities should indeed be considered to be disaster-prone to the extent that the infrastructural arrangements of urban society often represent high-risk technologies that are definitely not failure-free. Of course, this also holds for the air traffic above densely populated city districts – the more so when there are take-off and landing operations at an international airport in the close vicinity of a big city.

It is clear that modern city administration and its services should be well-prepared and well-equipped to cope with disasters in an effective way. Of course, if a nuclear attack hit a modern city, there would be almost nothing the city authorities and the urban services could do. But with other disasters, the city administration and its agencies should be able to manage the disaster. They should be able to take the appropriate emergency measures themselves and to engage in adequate co-ordination. Modern urban government should evolve self-sufficient tools to manage the immediate impact of a wide range of disasters. There may, nevertheless, be various reasons for resorting to assistance from outside.

The Amsterdam experience

In the emergency phase following the crash of the Israeli Boeing 747, counter-disaster, emergency relief and rescue measures, and on-the-spot operational co-ordination, were established quickly and efficiently. The overall advance preparations for major incidents and disasters in Amsterdam bore fruit here – and it is important to emphasise that this implied much more than simply following the letter of a disaster contingency plan. The chaos during the initial hours was to a large extent unavoidable; immediate availability of better communications equipment could only marginally have tempered but not eliminated the chaos.

The fact that the immediate emergency response was relatively effective and efficient was in part due to the location of the accident. The Bijlmer disaster hit the largest city in the Netherlands, with an appropriately sized and – so it transpired – professional administrative organisation. The region also has a large pool of private and volunteer bodies which – if properly managed and deployed – can make an important contribution in a support role. This fortunate situation endows the city authorities with greater flexibility of action. In some other municipalities, a Bijlmer-scale disaster might well create major shortfalls in capacity, and require a quite different approach. Following the immediate response, city authorities and services were confronted, often unexpectedly, with a variety of complex and politically sensitive challenges concerning entitlements to relief and support, and the position of illegal immigrants. Let us review briefly the performance of some of the principal municipal services involved in dealing with either or both phases of the disaster response:

- The *Amsterdam fire brigade* successfully withstood the pressure of events. Given the specific circumstances it was unable to fulfil its primary task of rescue and damage limitation. Once the fire had been doused its main role involved the unfamiliar and, in this case, highly indeterminate task of large-scale recovery of remains. Co-ordinating this was no easy job but, on the whole and given the immense time pressures superimposed on the operation, it proceeded satisfactorily.
- *The Amsterdam police* were confronted by a comprehensive and diffuse set of duties. The investigative operation to help establish the death toll and trace missing persons was a striking example. During the second week, the large number of illegal immigrants among those affected put the police in a delicate position. As the agency responsible in normal circumstances for identifying and expelling illegal immigrants, the police distanced themselves from the implementation of the *ad hoc* 'legalisation' procedure. They were understandably reluctant to become involved in a conflict of interests. However, this put responsibility for solving the problem of the illegal immigrants involved in the disaster with other sections of the city administration that were not always well equipped to deal with it.
- During the Bijlmer disaster the *Social Services Department* too had to cope with a set of large-scale, complicated and fragmented problems. While having an excellent programme for minor catastrophes, the Department had no inkling of the consequences of a gigantic, Bijlmer-type disaster. Partly through its own doing, the Department was confronted by a mounting log-jam of problems. It was not always able to reduce these to manageable proportions and was unable adequately to fulfil its co-ordinating role in the field of practical help. Even so, primary services to victims, family and friends were generally delivered in a timely and adequate fashion.
- The *Population Register* is an example of a service that was confronted by the events of the disaster with unexpected, important and intricate tasks. In fulfilling these tasks, particularly the registration of candidates for the Kosto list,

the Population Register did not have an easy time moving from individualised, highly routine to more or less collective, fast-track and highly sensitive people processing.

- The Bijlmer disaster came at a time when the position and tasks of the recently established Neighbourhood Councils with regard to disaster situations had not yet been adequately defined. This helped create a number of performance shortfalls by the *Southeast Neighbourhood Council*'s administration. There was a very long delay before it established a meaningful presence in the municipal Co-ordination Centre. It also had to struggle to fulfil its assigned duty of co-ordinating the services at the Bijlmer Sports Centre.

Despite their variable tasks, modes of operation and capabilities, overall emergency communication was good within and between the services combatting the disaster. Information and communication on the provision of material and psycho-social services were, however, occasionally hampered by less than smooth contacts between services and the geographical dispersion of relief locations.

Implications

The Bijlmer disaster attests to the necessity for modern cities to take counter-disaster planning and preparation seriously. Modern city government should resist complacency with regard to personnel policies, training endeavours and equipment for disaster agencies such as the fire brigade, the police, and, if relevant, civil defence. Furthermore, the Bijlmer disaster demonstrates that, whenever significant numbers of people become uprooted or otherwise afflicted by a disaster or during its aftermath, this may have immediate consequences for agencies that do not particularly define their mission in terms of disaster and crisis management. Such 'disaster-relevant' but generally not 'disaster-prepared' agencies as the Social Services or ethnic associations should be well aware that, depending on the characteristics of a disaster or crisis, they may, to their own surprise, move towards a prominent, highly visible and therefore also politically vulnerable role in disaster management.[1]

In a way, this argument also holds in a much broader perspective. For instance, city authorities and neighbourhood officials alike would be well advised to check proposals for municipal decentralisation and neighbourhood democracy in relation to their resilience in disaster situations. It is questionable whether much is to be gained from decentralising disaster management to the district level, since this would only imply a future need for immediate assistance from the city level. But the issue of scale may also be reversed: one may question whether regional bodies and counter-disaster systems are always sufficiently knowledgeable about, or at least sensitive to, the intricate fabric of the local community to adeqately grasp and cope with its particular vulnerabilities and disaster sub-culture.[2]

Leadership

Conclusion 2: Assertive political and operational leadership is a key asset in counter-disaster operations and crisis management.

Political and administrative leaders are put to the test in disaster situations and other crises.[3] One can argue that they will indeed face a number of tests. Their first and most difficult task is to come up with an adequate definition of the situation and to derive a cogent strategy from this assessment of what has happened and what is going to happen. This task will usually be hampered significantly by a lack of relevant information. Paradoxically, in a crisis atmosphere proper leadership may be weakened by hypervigilance and, for that reason, ill-considered judgements of the situation. More positively, leadership in crisis discloses a sensitivity to changing situations which may very well ask for a redefinition of what the crisis is all about.[4]

Political and administrative leaders should, secondly, adopt a line of thought and a course of action cherishing the need for co-ordination. They should engage in a balanced mix between 'governance' and delegation of authority. Their leadership will manifest itself in resolve and determination, but this should not be mistaken for detailed supervision over operational activities.

Thirdly, political-administrative leadership should take into account the unleashing of extreme collective stress during crises.[5] Under such circumstances, political leaders and administrators who do not understand the emotive and symbolic dimensions of disaster and crisis management will be doomed to fail. To be sure, this is not the same as saying that only a soft-handed approach to the tensions and conflicts that may arise in the course of a crisis is appropriate. More important is that leaders take into account the symbolic dimensions of whatever they do or fail to do in a crisis situation, and its impact on public perceptions in a situation when their authority and legitimacy may quickly erode under critical questions about who is to blame for the lack of preparedness, warning failures and other allegations that are endemic features of community responses to calamity and crisis.[6]

Fourthly, counter-intuitive as it may seem, leadership in crisis definitely involves a clear and distinct strategy of crisis termination. More often than not, politicians and agency chiefs do an excellent job in keeping up the crisis momentum, while neglecting to prepare for the psychological and organisational intricacies of the road 'back to normalcy'. In part, this may be because they suffer from a collapsed time perspective under the extreme pace of events. Partly also, there may be a habituation process at work, indeed a form of 'disaster addiction'. But there may also be more self-serving reasons for prolonging the crisis phase and the emergency regime of policy-making and implementation that typically grants wide leeway to senior executives and puts legislative powers and oversight on the back burner.[7]

The Amsterdam experience

The experience of crisis decision making during the Bijlmer disaster testifies to the vital significance of a well-functioning Co-ordination Centre. The logistics, organisation and decision-making strategy evident in the Amsterdam Co-ordination Centre are close to exemplary. At the same time, one has to bear in mind that every disaster follows a different course and will make its own demands on the Co-ordination Centre team. This begins with its composition. For example, a calamity in the form of an epidemic would need a quite different mix of expertise, and a very different configuration in the first and second circles of organisations and executives involved, as in the the Bijlmer disaster. Organisations that are almost totally excluded from the picture during an air crash such as this may be in the limelight next time around when a different type of disaster strikes. Moreover, the effective performance of a Co-ordination Centre depends on unambiguous agreements on which individuals and services should be present – and which should not – and clear communication and co-ordination between those involved.

Disaster management can be seen as almost by definition an inter-organisational task, whereby there may be differing calls on city services, and varying pressures may arise in the relationship between the city government and the public. Being explicit in disaster planning about several essential requirements and tasks of crisis decision making during a disaster (e.g. the classification as a disaster situation, the establishment of a Co-ordination Centre, deciding on who should and should not be in the Co-ordination Centre, and ending of disaster status) ought to be helpful in creating clarity in the relationship between the authorities, which facilitates the exercise of political and administrative leadership required for a quick and well-co-ordinated response. An element of chaos is inevitable during the first few hours of a disaster, but its duration and harmful impact need to be constrained as much as possible. Hence, the speed and quality of raising the alarm, communication between the various operational services and their performance at the disaster site need to be treated as priority items in forward planning.

Once again, the Bijlmer disaster bore out the established maxim of international disaster research: frequently, inter-agency co-ordination is not merely the solution, but also an important if not the core problem in disaster and crisis management. In general, however, there was a good level of co-ordination between those city services that were adequately represented in the Co-ordination Centre.

There was a willingness to co-operate. The tight management of meeting sessions and the undisputed leadership position of the Mayor contributed to this. Effecting the agreements reached in the Co-ordination Centre, members of the Co-ordination Centre team made frequent visits to various locations. This enhanced their insights into the situation and enabled the crucial, on-the-spot, personal contact with people working in trying circumstances. At the same time, the co-ordination and implementation of relief was hampered by the differences in perspectives and *modi operandi* between some of the city services (specifically the Social Services and the Housing Departments), as well as between the various central city services and the Southeast Neighbourhood Council, where central and local positions sometimes clashed and a pre-established regime for communication and conflict resolution was lacking.

As far as the co-ordination between the local and supra-local levels of administration was concerned, counter-disaster operations in this case were almost completely managed by the Amsterdam city authorities. Consequently, there was only a minimal need for co-ordination with other municipalities and levels of government. In general, contacts were satisfactory. Good relations were established with the representatives of most governments whose nationals were involved in the disaster. An important exception here were the difficult relations with the government of the Netherlands Antilles.

At given moments during the disaster period, the Co-ordination Centre set time limits for completion of certain tasks (the operation to recover remains, release of lists). This had the positive effect of establishing overall clarity, which in turn helped limit harmful speculation and rumour-mongering. An implicit risk of setting time limits was the enormous pressure imposed on the services involved.

Implications

Two potential lessons can be drawn from the experience of disaster and crisis management in the Amsterdam case. Firstly, as stated, co-ordination is not only the solution, but also a core problem in a fair number of questions of disaster response and crisis management. Many crises unfold as a sequence of rather divergent situations with variable clusters of actors in core or peripheral positions and different configurations of problems becoming important and urgent. Each step in this sequence – which does not necessarily follow the kind of fixed script implied by most textbook accounts of crisis management – tends to pose quite specific problems of co-ordination. For instance, the network of organisations and agencies dealing with the crisis may change considerably from one phase to the other. This may cause considerable frictions between agencies feeling they are losing ground as the crisis evolves, and those that shift into more prominent positions. One could see this as a particular version of the problem how to get 'back to normalcy' and how to mobilise agencies that, in an initial stage, do not seem to be important enough to contribute their services to the common cause.

Secondly, co-ordination needs and inter-governmental and inter-agency relationships are not only contingent upon the nature and scale of the disaster and the pre-existing administrative system, they are also a matter of perceptions and reputations. We do not intend to play down the importance of scale: where there are multiple disaster locations – Amsterdam being just one among them – the relationship and harmonisation between local and supra-local authorities would be quite different, at least in the Dutch context. In those cases, a mere decree from one Mayor would not be enough to make things happen, and there would need to be adjustments in line with quite different relations, politically and legally. It should be acknowledged that a country such as the Netherlands only has limited possibilities for supra-local political and administrative co-ordination. The Bijlmer case bears, for all its misery, on a limited disaster. The crisis that followed was, admittedly, more complex. But it generally remained within the purview of the city administration and was manageable from the perspective of inter-organisational co-ordination.

However, the fact that in this case local authorities acted almost independently of higher levels of government cannot simply be explained by pointing to the fact that it involved a disaster on a fixed location well within city limits. There had been precedents of other essentially local crises in Amsterdam where central government officials and agencies became much more strongly involved, on one occasion actually side-stepping the Mayor and other city officials and placing Amsterdam under a temporary form of direct rule. In the Bijlmer case, the reverse was true, and this was so partly because of the strong reputation of the Mayor and his assertive style of leadership right from the outset of the crisis. Virtually from the outset, he and his colleagues in the Co-ordination Centre were seen to be in charge and on top of things.

The social dimension of 'technological' accidents

Conclusion 3: Disaster managers should avoid a technological bias in designing and implementing counter-disaster strategies for urban environments. No matter what technological system triggers them, once they are manifest, the essence of urban disasters is the collective uncertainty and disruption they set in motion.

Major calamities demand more than the organisation and co-ordination of emergency rescue and relief. Alongside measures to combat the disaster, they also increasingly need crisis management: the regulation of the complex social problems that tend to unfold and reveal themselves immediately or more gradually in the aftermath of a disaster. Certainly, in a large-scale urban situation like Amsterdam, the decision-makers need to give this sort of crisis management high priority. Invariably, and indeed virtually by definition, major disasters are not confined to putting out fires, keeping the peace, maintaining security and mitigating the effects of toxic fumes. Hence a flexible outlook on the significance of the various city and other services to disaster and crisis management is required. As implied above, at certain stages of the crisis, city departments like Social Services, Housing and the Population Register will be far more important than operational services typically associated with disaster management like the fire service or police. Most specifically, adequate contingency planning for major urban incidents and disasters should anticipate the role of services dealing explicitly with the social dimension of crises: immaterial and material support, counselling, and after-care.

The Amsterdam experience

In this context, one can characterise the first two weeks of the Bijlmer air crash disaster as a series of evolving definitions of the situation: the hope that there might still be people who could be saved; emergency relief; establishment of casualty figures; co-ordination of relief; the problem of fraud by pseudo-victims; the issue of illegal immigrants in the affected area; the issue of illegals in and around Amsterdam. Some authorities and services found it difficult to cope with these dynamic processes. They had a difficult time switching from acute disaster response to more generic forms of social crisis management, yet, on the whole, they made every attempt to keep in tune with and help influence the evolving community perceptions of the situation.

The social dimension of disaster management begins right at the first moment. For example, disasters always spur a so-called 'mass assault' into the immediate area and the command/decision centres. With the Bijlmer disaster this collective behaviour phenomenon was well recognised and managed. The mass assault (volunteers, uninvited advisers, inquisitive onlookers) was brought under control quickly and efficiently. After the symbolically significant visit by the Queen and Prime Minister to the disaster area, the further influx of VIPs' (the 'elite assault') was firmly halted. So too the 'media assault' (journalists) was regulated effectively after some initial ups-and-downs.

One indication of the predominance of technological if not technocratic perspectives on disaster management is that most preparations and planning for major incidents and disasters pay inadequate attention to the whole question of victim relief.[8] When disaster strikes, it brings with it acute social problems related to organisation and co-ordination of victim relief. This was also true of the Bijlmer disaster, and the Amsterdam authorities were hard pressed to make up for the lack of pre-disaster relief planning. The problems were exacerbated by the multi-cultural background of those involved and political and administrative complications between City Hall and the Southeast neighbourhood administration. A strong point was that, from the outset, representatives of relevant victim organisations were given an important role in relief and after-care planning set in motion soon after the disaster had happened.

The preceding observations emphasise the point that effective leadership during disaster situations goes way beyond internal and external co-ordination. Paying due attention to the social dimension of urban disasters also involves making and communicating strategic policy choices which direct and constrain the operations. In particular, setting a clear course in regard to the emotional content of a major disaster is highly significant for the authorities' credibility and effectiveness. The Co-ordination Centre at City Hall took a caring attitude to relief and aid activities. Examples of this were equal treatment for victims regardless of immigration status, the linkage of material and intangible aid, the major focus on collective mourning and the speedy preparations for longer-term after-care. This emphasis on a humane, integral approach was a major factor contributing to the effectiveness and legitimacy of local governmental action in relation to the Bijlmer disaster.

From the second week onwards, the Bijlmer disaster was marked by the issue of the illegal immigrants involved, and the pseudo-victims attempting to 'hitch a ride'. Doing everything to avoid any extra burden on the illegals due to the disaster, fitted in with the concept of 'caring government'. The problem here was that the various city services applied differing terms and criteria when granting relief to victims. Therefore, the doctrine of 'caring government' needs refining to cope with the need for strict differentiation between those entitled to assistance and relief and people seeking to abuse the situation. This can help to prevent escalating claims for government services, limit the scope for misuse of the services offered, and pre-empt possible feelings of injustice caused by ambiguous and constantly changing criteria and rules about who is entitled to what forms of aid and assistance. Also, it can greatly relieve the pressure on government agencies that normally provide various forms of care but that are less often called to perform their services under disaster conditions.

From the perspective of the social dimension of disaster management, another strong point in the Bijlmer air crash case was that the authorities were clearly emotionally involved. Awareness of this sincere involvement was an important underlying factor in public appreciation for their work. At the same time, this involvement imposed a heavy psychological burden on the people in charge. The emphasis on maximum care and the considerable efforts by all concerned to make good on this promise channelled the collective stress caused by the disaster in the local community. For example, the arrangement and organisation of the mourning ceremonies and the psychological support for those involved were evidence of great care and consideration on the part of the authorities. These efforts were widely appreciated as such, in contrast to the uproar about the policy concerning illegal immigrants and some of its unanticipated effects on the public mood surrounding the disaster and the solidarity with its victims.

Implications

The major lesson to be drawn from the Bijlmer case is nothing less than the need for a fundamental reorientation of pure and applied theories of crisis management. It may be true that high-risk technology is a basic feature of modern society and its urban sector, and that from time to time modern society and big cities inevitably face disasters instigated by a loss of control over this kind of technology.[9] But the basic fact to understand is that technological risks may be at the beginning of a chain of events that may end up in a dazzling variety of crises. It would be quite wrong to assume that there would ever be one mechanical, if not manageable, psychological and social response to what may initially look like a quasi-simple disruption of the technological process. A totally different perspective on disaster and crisis management is in order. It asks for a comprehensive view of possible stakeholders and definitions of the situation in the wake of a disaster, spanning the entire continuum from a pre-occupation with the commercial losses inflicted by disaster-prone technology to the delegitimation of hitherto dominant forms organising urban economy and society. In the context of counter-disaster measures, it introduces a broad spectrum of groups, organisations and agencies which may carve out a voice and a role in the course of events.

In some limited ways, the Bijlmer air crash case takes us far beyond the notion of compound disaster.[10] It forewarns us of the complexities of modern urban society and the surprises that may await urban and national authorities who simply take an air crash at face value. Meanwhile, starting from the diversity of potential technological risks, this raises disturbing questions about the contingencies relating to modern urban society and the intricacies of mitigation and preparation practices concerning urban disasters and crises.[11]

Disasters in multi-cultural settings

Conclusion 4: Urban crisis managers have to reckon with the fact that, increasingly, major disruptions that occur in the city will affect ethnically diverse populations, confronting them with the need for a differentiated approach to warning, relief, communication, mourning and after-care.

It is common knowledge that minority groups differ among themselves and from majority citizens in terms of world view, socio-economic status, family organisation and structure, participation in religious and other voluntary associations, and in terms of political efficacy and trust in social and political institutions – but less so that they may behave differently in disasters.[12] At least, there has been a neglect of the differences in coping with disasters among ethnic groups and between them and the majority of the population. In a sense this is a remarkable missing link in our knowledge about disaster management. For there are several reasons why the ethnic factor could be a very important one in relation to disasters and crises. It adds a particular dimension to basic observations about effective warning, about the appropriateness of specific disaster responses, and about relief and recovery programms. But, more fundamentally, it lends further weight to competing questions concerning solidarity building and outbursts of social and political tensions. Thus disasters may confront multi-ethnic societies with the additional burden of how to overcome the built-in animosities and rivalries that often tend to dominate in the aftermath of a disaster.

The Amsterdam experience

The Bijlmer disaster affected various groupings of legal and illegal immigrants. Providing information to and communicating with immigrant groupings proved troublesome. Both practical limitations and fundamental dilemmas were involved here: a limited overview of the precise nature and scope of the groupings affected; language and cultural problems in the handling of information; and mistrust of the government, certainly when the climate of opinion surrounding the disaster became chillier. The Bijlmer disaster shows the specific problems that can be caused by a disaster in a multi-cultural community. The acknowledgement of these problems will be a lengthy process in the period ahead, all the more to the extent that disaster and crisis management has previously been a mainly 'white' affair.

Implications

It should be clear that the Bijlmer air crash emphasises the need for an additional set of changes of orientation in disaster and crisis management beyond what has been proposed above. It demonstrates that we should redefine the key factors of disaster geography. A disaster may be of the kind to hit hard on a specific category of the population and for that reason may need an equally specific approach. To take this one step further, if the authorities did not grasp the relevance of this perspective, they could be perfect disaster technocrats but would utterly fail in crisis management. The Amsterdam case is a very mild version of a 'multi-coloured' scenario, with the authorities intuitively (for example, not through systematic anticipation and planning) grasping the specific context of a disaster hitting an multi-ethnic district in a metropolitan area, yet having serious trouble in coming to grips with the ramifications of the event in and outside the particular urban setting. It is not too much to say that, in this respect, disaster and crisis management asks for a truly intimate knowledge of multi-ethnic mind-sets.

Otherwise, there could be a great many surprises beyond that of the disaster itself.

Another implication is that one should do away with the long tradition of disaster and crisis management as a predominantly 'white' affair. This goes back to those days when disaster and crisis management were part of 'white machismo'. Those days may be over now and 'representative bureaucracy' also needs to find its way to disaster and disaster-relevant agencies and organisations. But there is still a long way to go before we arrive at an adequate amount and quality of multi-ethnic 'thinking and doing' in disaster and crisis management.

Public information and media relations

Conclusion 5: Effective public information policies can only succeed if crisis managers develop and implement a comprehensive, consistent and attentive approach to the mass media. This compels crisis managers to develop a realistic perspective on the pivotal importance and the needs and behaviour of mass media in contemporary urban disaster situations.

Government officials and members of operational services often share negative views of the role of the media during disasters. They emphasise the media's disposition for sensational news. They deplore the way in which the media seem to be insatiable in their thirst for pictures and one-liners from victims, spectators and disaster tourists. They may have had personal experience of irresponsible reporting practices. They may have witnessed media crews roaming about a remote disaster-stricken community, claiming or paying dearly for scarce logistical facilities before the out-of-town authorities can. They may have heard the stories about media actions that gravely complicate crisis operations, for example during terrorist incidents and kidnappings. And, when they are candid, they may confess to anticipating with some anxiety the moment during some future crisis that they themselves will be pressed by the networks for information on the latest developments.[13]

These attitudes are beginning to change, not only among authorities but also among disaster researchers and media experts. There is a growing awareness that the media may indeed perform vital functions and that it is plainly impossible for the authorities and disaster services to cope with disasters and crises without media involvement. How to inform the public on potential threats? How to warn people of impending dangers? How to assure people that the government is taking proper action? How to engage in effective rumour control? In the absence of the mass media the authorities would not be able to accomplish these essential tasks as comprehensively and quickly as is often needed. Police and criminal justice officials have learned this lesson in other domains: missing persons or unsolved crimes shows make for prime-time television, and in some instances prove highly useful in getting to the bottom of seemingly hopeless cases. Hence the emerging view of the role of the media during disasters and crises is a more paradoxical one: yes, they can cause a lot of trouble and consume a lot of attention; but yes, they can also be tremendously helpful to communicate with the public.

Of course, to a very large extent contemporary disasters and crises have become media events. As the media version of the well-known sociological Thomas theorem has it, if the media define a situation in crisis terms, it will be a crisis in its consequences.[14] Taking a sober stance on this point, it is clear that, to a certain degree, effective crisis management will imply effective media management. By definition also, major disasters are international media events. This implies internationally oriented media management, including interpreters and adequate foresight on the pace, reporting styles and information needs of the international news media.

The Amsterdam experience

There was effective media management in and from the Co-ordination Centre, based on a strongly centralised model with the leading role for the city's Information Department. Essential to the model was the imposition of a fixed rhythm for agency briefings on the Co-ordination Centre team and subsequent decision making in the Centre, which became attuned to the twice-daily press conferences. It was a clear example of a conscious strategic decision early on in the disaster period to adjust the organisation and timing of crisis management to the pulse of the media and the need to get through to the public frequently and clearly. The unscheduled press conference at the Justice Ministry in The Hague, on the issue of the 'illegals' among the victims, slipped through this arrangement, and instantly caused a certain amount of confusion and uncertainty back in Amsterdam.

Implications

One can very well argue that the media have become chief stakeholders in (the management of) disasters and crises. Instead of being confined to their self-proclaimed role as mere outside observers, modern mass media are in fact taking a much more active part in the crisis process, and by their reporting often exert substantial influence on the course of events. This calls for a sensible approach on the part of the authorities towards the media, going beyond simplistic derogation of its 'irresponsible' behaviour. In fact, there is much to be said for the proposition that authorities tend to get the kind of media treatment they deserve: aloofness and defensiveness breed aggressive reporting, while a more pro-active and open approach increase the chances of solliciting a certain measure of co-operation at critical junctures. Consultation with media representatives on working practices – particularly of photographers and TV crews – in the disaster areas and relief/aid sites is as important as ever. This consultation process should be an integral part of the disaster planning process, and not be postponed until a disaster has happened, when time is lacking and tempers are short. This kind of consultation should not only deal with the regime of press conferences and site inspections, it should more generally cover the whole range of relations between journalists and operational services at the disaster site. For the media representatives involved in such a venture, such arrangements should be an integral part of quality control.

Some additional issues

Improving information processing and reducing uncertainty

Initially, information on the Bijlmer disaster concentrated on the number of dead and injured. It later transpired that the preliminary estimates of the death toll (250) were unfounded. Unduly impressed by the large number of missing person reports in the first few days, some authorities stuck with the inflated estimates for too long. Information that could have corrected these estimates at an earlier stage was not adequately processed into the Co-ordination Centre. In part, this was due to organisational problems, but there were also distinct biassing effects at work among key members of the Centre.

Such biasses can be mitigated if not entirely overcome by a number of devices. At key moments, bringing in outside experts to offer second or third opinions on essential but indeterminate or disputed elements can be of great help in the acquisition and verification of information. This should preferrably be done in a discreet and low-key manner, since there is always a somewhat delicate situation with regard to the internal experts and operational agencies who may feel they are not being taken seriously. In addition, given the major role played by rumour in highly uncertain situations, it is well worth establishing a rumour control system. From the very start, incoming information should be systematically checked, and never taken at face value. Staff agencies in co-operation with the Information Department should play a vital role in this screening process.

Stress management among crisis decision-makers

In the Bijlmer case, some authorities had to be alert to the risk of disaster addiction. The key decision-makers were all more or less locked away in a set of rooms underneath City Hall, captivated by round-the-clock television pictures of the disaster site and the recovery operations, and by the usual flood of agency briefings, incoming messages, inquisitive visitors and rumours. Some members of the Co-ordination Centre team took deliberate action to avoid this so-called 'bunker fever' by leaving the centre as much as possible for location visits. The accumulation of tasks and roles onto the Mayor during the Bijlmer disaster did not cause problems this time around, but it could very well cause severe physiological and psychological strains in the event of a more drawn-out disaster scenario. Once again, the Bijlmer disaster showed that some people in authority overestimate their psychological resilience under heavy pressure. It is worth considering stringent rules to prevent them from overexerting themselves: a strict working rota; limits on the number of roles per individual; reduction of the irreplaceability of executives; medical and psychological supervision on call; and careful evaluation of the psychological effects of continuous TV pictures from a disaster site.

From emergency response to strategic crisis management

In the hectic situation in the Co-ordination Centre in the first few days, it was not easy to devote sufficient attention to problems and considerations that lay beyond the horizon of the immediate disaster period – the drafting of the after-care plan being an important and fortunate exception. At the same time, our survey of the Amsterdam population indicates that the community was very sceptical about any long-term positive changes resulting from the Bijlmer disaster: air-traffic safety over Schiphol airport, the living environment in the Bijlmer and the position of illegal immigrants in this area were all expected to get worse rather than better. In addition, the Bijlmer disaster has exposed and dramatised a number of generic problems in a contemporary multi-cultural city. Governments – both local and national – will need to formulate answers to these. In the Amsterdam situation, the most marked concerned the situation of illegal immigrants (also outside the Bijlmer area); the quality of population registers; and the relationship between City Hall and the Neighbourhood Councils. Urban authorities must realise that, sooner or later, expectations or feelings of anxiety and injustice triggered or reactivated by a disaster will meet up with the memory of the 'caring government' doctrine promulgated during the disaster period.

Learning from the Bijlmer disaster?

Learning from peak events and crises is even more difficult than learning in public policy in general.[15] It is easy to overreact to a highly publicised policy failure, and it is as easy to be lulled into premature routinisation of *ad hoc* measures and improvised decisions that turned out to be helpful in one particular instance. Finding out what exactly happened and why, and drawing the right – if any – conditional generalisations about them for future use, is a delicate process in a post-crisis environment that is often characterised by blaming, scapegoating and an institutional battle for the dominant interpretation of events.[16]

Although generally perceived to have been managed effectively and therefore less a subject of official defensiveness and political contention, the Bijlmer air crash disaster is no exception to this general observation. It is not easy to learn the right lessons from this episode. For one thing, it would be tricky to regard the Bijlmer crash as a prototype of disasters and crises to come. In fact, one should not underestimate the fact that for all its misery the disaster could have been much worse – from both a victims' and a managerial perspective. The disaster meant instant death for more than 30 people, but only a few people were wounded. This made for a very peculiar kind of situation where, during the first hours of the disaster, almost nothing could be done. Several emergency services – first aid, hospitals – were mobilised but had little to do, a unique event and not a good guide to medical disaster planning. Within the context of learning, the Bijlmer disaster and the subsequent crisis bear witness to the importance of the social-psychological dimensions and to the need for public authorities to find effective answers to the collective stress that may affect urban society when it is confronted with a major disaster. To some extent, disaster response and crisis management during the Bijlmer air crash episode emulated the management practices after the Lockerbie tragedy in emphasising the notion of 'caring government'. For all the differences between these two events, this turned out to work well. But again this solicits questions about learning. Sooner or later people may learn all too well from this kind of policy philosophy and may expect the government to be overly generous in its relief and recovery programms.

The fact that in this case the management problems triggered by the original disaster were soon overshadowed by the social crisis that came to the fore in its wake implies another important lesson. In a sense one should be aware that 'the disaster after the disaster' may very well be quite different from the nature of the original misfortune, substantively as well as socially and politically.[17] Contrary to the initial idea of 'disaster after the disaster' which simply points to the later phases in a cyclical process of disaster management, the Bijlmer case shows that a disaster may unleash forces which indeed may result in crises of a totally different nature – going all the way from an air crash in an urban district to multi-ethnic turmoil in urban society at large.

As a result, the final and possibly most important lesson to be drawn from this case is that, while most people will remember the Bijlmer air crash disaster as an instance of effective and highly successful disaster management, political and administrative realities were more complex and ambivalent – with a fair share of successful as well as less successful probes and practices. In other words, the now dominant image of a model-like example of effective disaster response is likely to become modified in due course as people will look back at the events from different vantage points. In fact, each major crisis goes through waves of success and failure in the eyes of history.[18] When this happens, the lessons of the Bijlmer air crash for urban crisis management may become what they should be: a topic for reflection and debate, a set of options with pros and cons attached to them, rather than a shortlist of mechanistic do's and don'ts.

Notes

1 See R.R. Dynes (1974). *Organized Behavior in Disaster*, Lexington-Heath, Columbus.
2 A.H. Barton (1970). *Communities in Disaster*, Doubleday, Garden City. Compare, for example, discussions of the US, French and British situation in: P.H. Rossi, J.D. Wright, and E. Weber-Burdin (1982). *Natural Hazards and Public Choice: The State and Local Politics of Hazard Mitigation*, Academic Press, Academic Press; C. Gilbert (1990). *La Catastrophe, l'Élu et le Préfet* (1990), Presse Universitaire de Grenoble, Grenoble; D. Parker and J.W. Handmer (Eds) (1991). *Hazard Management and Emergency Planning: Perspectives on Britain*, James & James, London.
3 See Y. Dror (1988). Decisionmaking under disaster conditions, in: L.K. Comfort (Ed.), *Managing Disaster: Strategies and Policy Perspectives*, Duke University Press, Durham.
4 I.L. Janis and L. Mann (1977). *Decision Making: A Psychological Study of Conflict, Choice, and Commitment*, Free Press, New York.
5 A.H. Barton, op. cit.; E.L. Quarantelli (Ed.) (1980). *Disasters: Theory and Research*, Sage, London.
6 M. Douglas (1992). *Risk and Blame: Essays in Cultural Theory*, Routledge, London; P. 't Hart (1993). Symbols, rituals, and power: the lost dimension of crisis management, *Journal of Contingencies and Crisis Management*, Vol. 1, 1, 37–51.
7 Compare J. Roberts (1988). *Decision-making in International Crises*, Macmillan, London, with M. Edelman (1977). *Political Language: Words that Succeed and Policies that Fail*, Chapter 4, Academic Press, New York.
8 R. Gist and B. Lubin (Eds) (1991). *Psychosocial Aspects of Disaster*, Wiley, New York; P. Hodgkinson and M. Stewart (1991). *Coping With Catastrophe: A Handbook of Disaster Management*, Routledge, London.
9 As argued, for example, by C. Perrow (1984). *Normal Accidents: Living with High Risk Technologies*, Free Press, New York.
10 R.R. Dynes, op. cit.
11 See, for example, the rather limited range of situations dealt with in R. Sylves and W. Waugh (Eds) (1991). *Cities and Disaster*, Charles C. Thomas, Springfield.
12 R.W. Perry and A.W. Mushkatel (1984), *Disaster Management: Warning, Response, and Community Relocation*, Quorum, Westport, 5; see also, by the same authors (1985). *Disaster Response and Preparedness Among Minority Citizens*, Georgia University Press, Georgia.
13 See L.M. Wilkins, L. Wilkins and T. Walters (Eds) (1989). *Bad Tidings: Communication and Catastrophe*, Erlbaum, Hillsdale.
14 W. Thomas and I. Thomas (1927). *The Child in America*, Chicago.
15 Compare L.S. Etheredge (1981). Government learning: An Overview, in S. Long (Ed.), *Handbook of Political Behavior*, Plenum Press, New York, 73–161; R. Neustadt and E.R. May (1986). *Thinking in Time: The Uses of History for Policy Makers*, Free Press, New York; P. Tetlock and G. Breslauer (1990), *Learning in American-Soviet Relations*, Westview, Boulder; R. Rose (1993). *Lesson-Drawing in Public Policy: A Guide to Learning across Time and Space*, Chatham House, Chatham.
16 U. Rosenthal, P. 't Hart and A. Kouzmin (1991). The Bureau-politics of Crisis Management, *Public Administration*, Vol. 67, 2, 211–233.
17 Kai Erickson (1976). *Everything in its Path: Destruction of Community in the Buffalo Creek Flood*, Simon & Schuster, New York.
18 Even seemingly clear-cut cases such as the Munich crisis, see, for example, M Latynski (1992). *Reappraising the Munich Pact: Continental Perspectives*, Johns Hopkins University Press, Baltimore and Woodrow Wilson Centre for International Scholars, Washington.

APPENDIX 1

Chronology of the Bijlmer Disaster

This chronology provides a concise picture of events involved in the Bijlmer disaster between the period of Sunday evening 4 October 1992 up to and including the withdrawal of the disaster declaration on Tuesday 27 October 1992. The chronology is largely based on the co-ordination Centre log book. Several other log books and evaluation reports of city departments and services have also been used.

Sunday 4th October

18.36 An El Al Israel Airline freighter (Boeing 747-200F) crashes on the 'Groeneveen' and 'Kruitberg' apartment buildings in the Bijlmer; 31 flat units collapse and 49 are totally burned out.
The first 06-11 emergency phone calls are received by the Central Ambulance Station, to the effect that an aircraft has crashed on the Bijlmer. The precise site of the accident is still not certain at this point in time.

18.39 Police duty commander informs the central incident room that a commercial aircraft has crashed into the 'Groeneveen' apartment building, which is now on fire.

18.43 Head of Information, Mrs Van Oostveen, informs Mayor Van Thijn about a possible crash in the Bijlmer.

18.44 Police officers are present at the disaster site. There is a gigantic blaze and the situation is very confused.

18.47 Chief of Police Nordholt warns Mayor that an aircraft has crashed on the Bijlmer.

18.48 First ambulance arrives at the site.

18.51 Mayor makes a verbal declaration of a disaster situation and heads for the Co-ordination Centre. The Chief of Police is requested to go to the Co-ordination Centre. Administration and Legal Affairs staff make the Co-ordination Centre operational.

19.10	Mayor, Chief of Police arrive at the Co-ordination Centre.
19.30	In the Co-ordination Centre the decision is taken that city information on the disaster will be co-ordinated from the City Hall.
20.00	The following people are present in the Co-ordination Centre: Mayor, Chief of Police, Fire Chief, Medical Services Deputy Director Coutinho, Social Services Director Van Dijk.
20.06	Police command post is established at Flierbosdreef police station.
20.20	Fire Chief reports in the Co-ordination Centre that there have been eight deaths. He also reports that 47 flats have been hit, and that an estimated 150 people were present in the affected buildings.
20.32	National Police Amsterdam Incident Room receives renewed reports of enormous crowds in the Bijlmer. Reports say that disaster tourists are breaking into police cars.
evening	Police spokesperson announces that there are probably 150 dead.
20.46	Fire Chief reports in the Co-ordination Centre that all relief organisations have been mobilised.
20.50	Population Register makes a print-out of the number of registered residents in the 'Groeneveen' and 'Kruitberg' apartment buildings. On the orders of the Mayor the Co-ordination Centre breaks into all channels of the Amsterdam cable network. Every four minutes, people are asked to phone with information.
21.07	The Fire Brigade announces that the fire is under control.
21.15	Medical Services reports to the Co-ordination Centre that 12 people have died.
21.56	Fire Chief reports in the Co-ordination Centre that 120 flat units have been destroyed and that no survivors are likely to be found under the wreckage. There are sufficient ambulances in position but there are only a few wounded.
22.00	City Hall Press and Information Department holds a press conference.
22.30	Mayor, Chief of Police and Fire Chief leave the Co-ordination Centre to visit the disaster site.
22.50	The Fire Chief reports to the Co-ordination Centre that the aircraft has bored into the 'Kruitberg' apartment building, the impact being five flats across and 10 high. He says that all the flats in this section have collapsed and that this is 'a large scale disaster'. According to his latest information, 12 dead have been removed. The Co-ordination Centre is now aware that the scope of the disaster covers a total of 80 destroyed flat units with 239 registered residents.
23.30	Mayor, Chief of Police and Fire Chief return from their inspection of the disaster site.

In the Co-ordination Centre the Bijlmer Sports Centre is designated as the central relief location for the homeless and victims.

Monday 5 October

00.00 Mayor, Fire Chief and Chief of Police give the first joint press conference.

01.15 Briefing Co-ordination Centre:
First reports of looting reach the Co-ordination Centre.
The Medical Services reports that the claim that 12 bodies have been recovered is incorrect.

03.56 First body is recovered.

05.48 Number of bodies recovered: 2.

06.10 Briefing Co-ordination Centre:
The Fire Brigade start the recovery operations.
Social Services reports at the Co-ordination Centre that 209 persons have been reported missing.

06.26 Number of bodies recovered: 3.

07.10 Number of bodies recovered: 5.

07.45-09.05 Mayor reports the situation to various ministers and government officials: the Prime Minister, his Deputy, the Minister of the Interior and the Director-General of Public Order and Safety.

10.40 Fire Chief reports to the Co-ordination Centre the rumour that there was a disco party in the affected flats. Neither the Southeast Neighbourhood Council nor Administration and Legal Affairs can confirm the rumour.

11.20 Briefing Co-ordination Centre:
Social Services crisis team at the Vlaardingenlaan is in charge of setting up relief facilities for the longer term.
Co-ordination of financial assistance, including alternative accommodation, is entrusted to Social Services. Social Services will represent city Housing Department at the Co-ordination Centre. City Housing Department will arrange the implementation of the rehousing operation together with the Nieuw Amsterdam Housing Co-operative.
Hot-line is installed between the offices of the Southeast Neighbourhood Council and the Co-ordination Centre.

11.25 Number of bodies recovered: 6.

11.30-12.40 Queen Beatrix, Crown Prince Willem-Alexander, a cabinet delegation led by Prime Minister Lubbers and Provincial Governor Van Kemenade, visit the Sports Centre and the site of the disaster accompanied by a delegation of the Municipality of Amsterdam.

12.40	Briefing Co-ordination Centre: It is decided at the Co-ordination Centre not to distinguish between those legally and illegally resident in the Netherlands when providing relief. Disaster Identification Team arrives at 'Groeneveen'.
afternoon	There is a meeting about rehousing the victims. Those present include Alderman Genet, the Chairman of the Southeast Neighbourhood Council, the Director of the Nieuw Amsterdam Housing Co-operative and the Director of the city Housing Department. City Housing Department is informed by Alderman Genet about the agreement that Social Services will handle the overall co-ordination of the provision of material relief and represent the city Housing Department at the Co-ordination Centre. Genet is informed about the situation with respect to rehousing the victims. Police set up an investigation team to investigate the number of missing persons.
13.25	Briefing Co-ordination Centre: Risk of collapse of the affected flats is still too high to permit the large-scale commitment of recovery crews. Police report the following information to the Co-ordination Centre: Number of missing persons: 209. Southeast Neighbourhood Council is entrusted with co-ordination at the Sports Centre.
15.25	British Consul General and the Dunedin group offer assistance. This could include sending experts who assisted with the Lockerbie disaster in Scotland.
16.00	Extraordinary meeting of the City Council to discuss the Bijlmer disaster. At this meeting the Mayor expresses fear that as many as 250 may have lost their lives in the disaster.
16.40	Briefing Co-ordination Centre: No further bodies have been recovered. It is decided at the Co-ordination Centre that the press will be admitted to the site of the disaster. There is still no reliable information at the Co-ordination Centre about the rumour of a disco party in the common areas under the flats. The Southeast Neighbourhood Council reports to the Co-ordination Centre that 600 people have been reported missing.
20.10	Number of bodies recovered: 7.
20.30	Minor fire occurs in the affected flats.
22.35	Number of bodies recovered: 8.
23.49	It is decided at the Co-ordination Centre to appeal to illegal aliens, via the press, to report if they have any information about missing

persons. They are assured that this will have no adverse consequences for them.

Tuesday 6 October

00.36 Number of bodies recovered: 9.

00.48 Number of bodies recovered: 10.

02.12 Number of bodies recovered: 11.

02.42 The 'Groeneveen' block has been searched completely. All but three apartments in the 'Kruitberg' block have been searched. No victims were found. Some 30–33 apartments cannot yet be entered.

06.00 Briefing Co-ordination Centre:
Mayor wants to use special equipment to check through the rubble whether there is a large number of victims in the common spaces. Fire Brigade check whether this equipment is available. There is still no certainty about the rumour of a disco party in the basement of the affected flats.
There are discussions at the Co-ordination Centre about speeding up the recovery operations. The careful approach taken by the Disaster Identification Team is too time-consuming. It is decided to speed up the recovery operations. Recovery operations should be completed within four days.
Number of bodies recovered: 12

10.20 Briefing Co-ordination Centre:
The recovery operations have been speeded up.

11.30 Prime Minister Mrs Liberia-Peters of the Netherlands Antilles arrives at Schiphol.

11.45 Briefing Co-ordination Centre:
The recovery operations are proceeding rapidly. Few victims have been found. Twenty police officers are staffing the phone lines to deal with calls about missing persons.
Recovery operations should be completed at midnight on Friday.
Permanent psychological assistance is provided on site to the police and the Fire Brigade.
Alderman Wildekamp will draw up an action plan for the provision of long-term relief.
Social Services reports to the Co-ordination Centre that rehousing is proceeding well.

13.30 Southeast Neighbourhood Council reports problems regarding coordination at the Sports Centre and organisation of overnight accommodation.

13.55 Prime Minister of the Netherlands Antilles and her officials arrive at the Flierbosdreef police station.

14.35 At the Co-ordination Centre it is decided not to accommodate all the growing number of requests by officials to visit the site of the disaster, the Sports Centre and the Co-ordination Centre.

16.05 Number of bodies recovered: 13.

16.20 Briefing Co-ordination Centre:
Fire Chief reports 'disappointing results' in terms of the number of victims found. It is not possible to employ more equipment and personnel on the site of the disaster.
Number of bodies recovered: 14.
There is still uncertainty about co-ordination at the Sports Centre. The Co-ordination Centre gives the Chairman of the Southeast Neighbourhood Council responsibility for co-ordination at the Sports Centre.

17.15 Co-ordination Centre decides that the legal or illegal status of the victims will not be recorded.

17.30 Number of bodies recovered: 15.

18.20 Number of bodies recovered: 16.

19.10 Prime Minister of the Netherlands Antilles hands a list of missing Antilleans to the authorities for use, in the investigations into missing persons.

21.00 Police report at the Co-ordination Centre that over 1,000 people have been reported missing and add that this cannot be a realistic number.

Wednesday 7 October

04.24 Number of bodies recovered: 24.

05.37 Briefing Co-ordination Centre:
The recovery operations are on schedule.
The files and records used in the police investigations provide inconsistent information about the number of victims.

08.15 Number of bodies recovered: 25.
During the day, Housing representatives are surprised by the number of victims, which still amounts to 250 according to the Co-ordination Centre. Every other hour the city Housing Department faxes lists of the victims to the Co-ordination Centre.
Deputy Minister of Justice Kosto sends two letters about the status of illegal victims to the Mayor and the Chief of Police.

11.00 Number of bodies recovered: 32.

13.10 Visits by diplomats are reduced, both to the Co-ordination Centre and the disaster area. The large numbers of visitors interfere with operations. Fire Brigade are irritated by the visitors.

14.45 Social Services reports to the Co-ordination Centre that the number

of people requesting overnight accommodation is rising.
Social Services reports that more accommodation is needed because dozens of illegal aliens are reporting to the Sports Centre, which is already filled to capacity.

16.07 Lowest floor of the flats has been reached.

16.37 Briefing Co-ordination Centre:
Recovery operations are ahead of schedule.
Special team of police detectives is formed to verify who is missing.

21.00 The tense situation at the Sports Centre is discussed at the Co-ordination Centre.
The problems at the Sports Centre are due to the following factors:

- Social Services uses standard forms for the payment of an advance to those affected by the disaster. Incorrectly, these forms refer to repayments;
- the various religions are not considered at meals;
- a deposit of G 265 has to be paid when a new home is allocated.

21.40 Nieuw Amsterdam Housing Co-operative reports the current situation to the Co-ordination Centre:

- 213 households have applied for rehousing;
- 20 houses are ready, six houses are now occupied;
- Antilleans are occasionally aggressive to staff as they feel that they are discriminated against compared to whites.

Thursday 8 October

00.00 It is decided at the Co-ordination Centre that the Southeast Neighbourhood Council will set up a new system to issue security passes for the Sports Centre, in co-operation with the Police. The recovery operations are a day and a half ahead of schedule. It is not expected that many more victims will be found. The number of victims is unclear. The Fire Brigade and Medical Services suggest that the bodies of some victims might have been reduced to ashes completely. The Mayor wants to obtain information from internationally renowned experts.

08.13 Briefing Co-ordination Centre:
The National Co-ordination Centre is asked by the Co-ordination Centre to fly in international experts who can decide whether victims might indeed have been reduced fully to ashes.
Social Services reports that the relief facilities are attracting more and more people.

09.18 Pathologist Professor Busuttil and Deputy Chief Constable Orr, both of whom worked at the Lockerbie disaster, are travelling to the Netherlands.

10.30 Alderman Genet, Alderman De Grave and Director Denijs have a meeting at the city Housing Department crisis centre about the lists, provision of relief at the Sports Centre and the criteria for providing help to those affected. Housing has now been offered to almost all those affected.

Ministry of Welfare, Health and Cultural Affairs announces that additional funds will be made available for psychological support to the survivors of the Bijlmer disaster.

14.10 Professor Busuttil arrives at Schiphol and is welcomed by the Mayor.

14.28 Problems at the Sports Centre are increasing.

15.30 Professor Busuttil visits the disaster site.

16.06 Recovery operations have been stopped.

Number of bodies recovered: 50.

Professor Busuttil reports on the situation. He thinks that it is likely that 10% of the victims will never be found.

19.40 Provincial co-ordination centre is closed.

20.27 Briefing Co-ordination Centre:

The list of missing persons will be released at the Sports Centre, at the offices of the Southeast Neighbourhood Council and on cable television. Agreements have been made with the media.

23.00 Alderman Genet proposes that a group of representatives of Social Services, Housing, Nieuw Amsterdam Housing Co-operative, Medical Services and the police should meet to discuss the scale-down of the operations at the Sports Centre.

Friday 9 October

08.30 Briefing Co-ordination Centre:

Three lists will be published:

- List 1: identified victims;
- List 2: persons who were in the vicinity of the flats at the time of the disaster and who have almost certainly died;
- List 3: persons who have been reported missing and who live outside the disaster area.

15.25 Alderman Genet draws the attention of the Co-ordination Centre to two matters that he expects to receive considerable publicity:

- procedure used when allocating housing to illegal aliens;
- scale-down of operations at the Sports Centre. The action plan drawn up assumes that operations at the Sports Centre will be reduced from 12 October.

17.00 Briefing Co-ordination Centre:

Current numbers on the lists:

- List 1: 9;
- List 2: 48;
- List 3: 63.

17.35 The mood at the Sports Centre is quiet. About 1,000 people are present. There are three relief workers for each victim.

18.57 The lists of missing persons are released and issued to those concerned. The lists contain a total of 120 persons.

20.34 Briefing Co-ordination Centre:
The Co-ordination Centre decides to agree with the scale-down of the activities at the Sports Centre, starting Monday 12 October.
Social Services reports that there are 'freeriders' who use the facilities although they are not entitled to them.

evening Group of over 70 Antilleans is accommodated at the Park Lake Hotel in Amsterdam South after intervention by the Netherlands Antilles authorities.

Saturday 10 October

00.05 The lists contain errors.

01.04 The Co-ordination Centre decides that a new, revised list will be released in the course of the morning.

09.50 It is agreed that, starting from Monday, the Co-ordination Centre will largely be concerned with co-ordination and its staffing will be reduced.

10.50 The Social Services concludes a contract with the Novotel (a hotel), to provide accommodation for the victims, following the initiative by the Netherlands Antilles government.

Sunday 11 October

11.00-13.00 Memorial service in the Bijlmer and a silent procession to the site of the disaster.

15.00-17.00 Memorial service in the Amstelhal of the RAI exhibition complex. The service is attended by 12,500 people.

16.42 The police receive three reports of bombs at the RAI. The building is not evacuated.

Monday 12 October

08.10 Briefing Co-ordination Centre:
It is decided that the disaster declaration will not be revoked for the time being.

	The after-care plan is presented by Alderman Wildekamp. The plan aims to provide combined financial and psychological support. Discussions about the assistance provided at the Sports Centre. The Southeast Neighbourhood Council argues that the Sports Centre should remain open until the funeral service on Sunday 18 October. The opening times will be 08.00 to 18.00. The following will be present at the Sports Centre during the opening times: Regional Psychiatric Aid Agency, pastoral support and Medical Services. Housing and Social Services will be present for two hours every day.
11.00	Nieuwenhuis (secretary of the Southeast Neighbourhood Council) tells Director Denijs that Housing should be present at the Sports Centre. Initially, Denijs refuses. Later Alderman Wildekamp phones to propose a compromise. Denijs concedes: Housing will maintain a presence several hours daily at the Sports Centre for two more days.
11.15	Scaling-down of operations at the Sports Centre causes problems.
11.30	Housing no longer wants to be present at the Sports Centre.
22.24	The Novotel now accommodates 197 persons in 94 rooms.

Tuesday 13 October

09.12	Briefing Co-ordination Centre: There is a meeting between the Mayor, Aldermen Genet and Wildekamp and the Southeast Neighbourhood Council about the scale-down of operations at the Sports Centre.

Wednesday 14 October

09.31	Briefing Co-ordination Centre: There is a discussion about the problems with the registration of Ghanaians. Mayor and Chief of Police reiterate that legal or illegal status should not be relevant when providing support to victims or reporting missing persons. There are problems with separating the 'freeriders' from the real victims. A working party of the Southeast Neighbourhood Council, Population Register, Social Services, Housing and the police is set up to develop a proposal. It is decided that sanctions may be taken against profiteers at a later stage.
afternoon	Chief of Police and the Deputy Minister of Justice Kosto have an emergency meeting and agree that illegal aliens affected by the disaster may obtain legal status under certain conditions.

Thursday 15 October

09.12 Briefing Co-ordination Centre:
There is a great increase in 'victims' at the Novotel and the Park Lake Hotel. According to Social Services the numbers have risen explosively to about 400 persons.
It is expected that the disaster declaration will be revoked on Tuesday 20 October.

afternoon Special meeting of the Administration and Legal Affairs support committee which is also attended by the Mayor, Chief of Police and Fire Chief. The potential abuse of facilities provided by the city is a major topic during the meeting.

Friday 16 October

09.14 Briefing Co-ordination Centre:
The Co-ordination Centre decides to continue to operate with a skeleton staff.
The Mayor leaves for a four-day visit to Indonesia. He is replaced by Deputy Mayor De Grave.

Saturday 17 October

09.00-18.00 Illegal aliens can register for the Kosto list at an office of the Population Register at the Sports Centre. The number of people who report is higher than expected. It is not possible to register everyone.

Monday 19 October

09.06 Briefing Co-ordination Centre:
The Deputy Mayor and Administration and Legal Affairs stress that the time limit for registration for the Kosto list elapsed on Saturday 17 October at 18.00. This limit should be adhered to.
Housing requests a clear decision about the allocation of housing to illegal sub-tenants who cannot prove that they have been resident in the flats for more than one year. The Deputy Mayor stresses that all victims have to be helped.

11.00-19.00 The Bijlmer Sports Centre is closed for the support of victims.

Thursday 22 October

16.00 Briefing Co-ordination Centre:

Total number of bodies remains at 43. Another 285 or so illegal aliens have reported to the Population Register on Thursday. It appears that some have reported more than once. There are also rumours that encourage the influx of illegals. It is expected that some 100 illegal immigrants have actually been affected.
A provisional list of 52 names is sent to the Deputy Minister of Justice. More names may be added at a later stage.

Friday 23 October

13.15 Briefing Co-ordination Centre:
It is decided that the influx at the Population Register has to be stopped.
At 15.00 registration for the Kosto list is stopped.

Tuesday 27 October

The Mayor revokes the disaster declaration.

APPENDIX 2

Notes on the population survey

The survey aimed to provide insight into how the people of Amsterdam viewed the performance of the city authorities and services during the Bijlmer air disaster. To this end, respondents were asked about their knowledge of the city services and departments that were involved in dealing with the disaster and about their opinion on the performance of these institutions and of the city administration as a whole. In addition, the poll included a number of general questions about their knowledge of the disaster and about its long-term ramifications in the affected area. The survey was commissioned by the Crisis Research Center and carried out by the Intomart polling agency. This appendix explains the research method applied and presents a summary of the most significant results.

Research method

The data were collected by means of a standardised questionnaire conducted by telephone among residents of Amsterdam. Respondents were chosen at random from the PTT telephone directory. The questions were put to the first family member, aged 18 or over, whose birthday occurs earliest in the year. This was done in the period 17–19 March 1993, five months after the disaster. We obtained data from a total of 363 households. No response was received in 45% of all telephone calls, because people could not be reached or refused to take part for other reasons. The survey is representative of the population of Amsterdam in terms of both sex and age. Since one of our aims was to draw separate conclusions on those living in the affected neighbourhood, more respondents were chosen from Southeast Amsterdam than from any other district (see Table A2.1). All results relevant to the entire opinion poll have therefore been corrected to reflect the actual ratio between the residents of Amsterdam Southeast and the rest of Amsterdam (in accordance with population statistics supplied by the city Press, Information and Public Relations Department). Table A2.1 shows the composition of the opinion poll before and after ratio corrections.

Performance of the city authorities. Respondents were asked to name the city services and departments involved in dealing with the disaster. These services were not specifically named in the question's formulation: the answers were spontaneous. Amsterdam city police and fire brigade received the highest scores, followed by the city Medical Services Department. The Southeast Neighbourhood Council and the city Population Register emerged as much less visible services (see Table A2.2).

This result indicates that in the view of the public it is mainly the people in uniform who call the shots in countering the effects of a disaster. However, it is possible that the question's formulation (involvement in the air disaster) led respondents to think in terms of *acute* disaster relief operations, in which particularly the uniformed services are in evidence first. Perhaps the phrasing of the question focused less attention on activities during the later stages of the disaster period, such as rehousing those made homeless. It may also be that recollection was mainly determined by media pictures from the first week following the crash, when mainly the fire brigade and police featured prominently in media reports. The shift in the relative importance of the various services involved appears to have been markedly less apparent, or more easily forgotten.

TABLE A2.1 *Composition of the opinion poll.**

	% (unweighted)	% (weighted)
Man	51	50
Woman	49	50
Aged 18–34	43	45
Aged 35–49	28	26
Aged 50 and over	29	29
Dutch	89	90
Surinamese	5	4
Other nationalities	6	6
Amsterdam Southeast	32	13
Rest of Amsterdam	68	51

**n*=363.

Respondents were also asked to state their views on the performance of a number of specifically named services. The outcome is predominantly positive (see Table A2.2). Only the city Population Register appears to have paid for the after-effects of all the publicity surrounding the registration of illegal immigrants, and for the view that 'the city authorities don't know their own citizens', with a lower rating. The Amsterdam people also praised City Hall's performance. In the view of more than three-quarters of the respondents, the city authorities reacted well to the Bijlmer disaster; 17% of them even expressed great appreciation. The performance of Mayor Ed van Thijn was given a similar rate of approval. A minority (16%) indicated that the disaster had altered their general perception of the functioning of the city administration. The larger part of this group said it now took a more positive view of City Hall. The information provided by city officials was also generally seen as adequate. Regarding the performance assessment of the various city services and the city administration, no significant divergences were found between residents of Amsterdam Southeast and those living elsewhere in Amsterdam.

TABLE A2.2 *Visibility and assessment of city services and departments*

	Visibility (%)*	Rating*
Fire Brigade	78	4.4
Mayor	11 **	4.1
Population Register	5	2.7
Housing Department	11	3.3
Medical Services Department	44	4.3
Social Services Department	21	3.4
Police	78	4.0
Regional Psychiatric Aid Agency (RIAGG)	10	–
Southeast Neighbourhood Council	4	3.6
Amsterdam City Hall	–	4.1
	n^*=363	n =363

* Visibility is expressed in terms of the percentage of respondents who named a city service or department. Nine per cent failed to name a single institution. The table does not list the services mentioned by less than 5% of the respondents, such as the Sanitation Department and the Transport Department. The rating of the departments is measured on a 5-point scale, ranging from very good (5) to very bad (1).

** The fact that respondents were asked to name city services involved in dealing with the Bijlmer disaster may have negatively influenced the number of times the Mayor was mentioned.

Interpreting these results obviously requires a good deal of prudence. The poll was conducted several months after the disaster. It cannot be ruled out that the results partly reflect a more general appreciation for the authorities concerned. This may well be the case as a wide majority of the respondents (94%) did not suffer any damage themselves as a result of the disaster. It is therefore most likely that these respondents did not have any direct experience with the city's performance in dealing with the disaster. Moreover, the services and departments that were most visible also received the highest scores. As for the services showing lower ratings (Housing Department, Social Services Department, Population Register), the number of people that refrained from giving any assessment was considerably larger (average 23%) than in the case of the more highly rated and more visible services (average 10%). As many as 50% of the respondents had no opinion on the performance of the Southeast Neighbourhood Council. For that matter, residents of Amsterdam Southeast more often (67%) expressed a view on that Neighbourhood Council's performance than those living elsewhere in Amsterdam (48%).

Longer-term effects. In sharp contrast with the positive appreciation for the city's performance during the disaster, scepticism prevailed when it came to the expected effects of the air crash in the longer term. Respondents were asked whether the Bijlmer disaster would affect air-traffic safety around Schiphol international airport. Eighty-five per cent thought it would not. The result is underpinned by answers given to an open question, which allowed respondents to cite issues which they themselves deemed important. Thirty-one per cent of all respondents answered this question. Nearly a third of their remarks (31%) concerned air-traffic. By far the most comments (84%) expressed concern about the lack of air-traffic safety over residential areas and the inadequacy of existing measures to tackle this problem. A narrow minority played down the commotion surrounding the safety risks of flying over residential areas.

TABLE A2.3 *Long-term effects of the Bijlmer disaster.*

	Southeast Amsterdam (%)	Other districts (%)	Whole of Amsterdam (%)
Schiphol air-traffic around airport			
much safer	1	2	2
a little safer	17	10	11
not safer	78	86	85
don't know	4	3	3
Consequences for the quality of life in the Bijlmer			
considerable	38	47	46
negligible	22	24	24
none	37	27	29
don't know	4	2	2
	n=47	*n*=316	*n*=363

More than two-thirds of the respondents believed that the disaster would have implications for the general quality of life in the Bijlmer. It is striking that this expectation was registered less frequently among residents of Amsterdam Southeast than among local citizens living in other areas. Eighty-six per cent of the respondents who expected implications for quality of life believed that these would turn out to be negative. Remarks made by respondents answering the open question also addressed the quality of life in the Bijlmer. In addition to the above-mentioned comments on air-traffic, we find various remarks on the relations between foreign immigrants and native inhabitants. A negative result was projected on both issues. Regarding the relations between the various population groups, the prevailing tenor of the remarks was that too much attention had been given to foreign immigrants compared to Dutch people. Very few indeed noticed increased solidarity.

The results derived from the open question which allowed respondents to raise matters, call for a final comment. A large majority (75%) of the remarks made by the respondents concerned negative criticism of a wide range of issues, most of which have been dealt with here. It is possible that those respondents who voiced dissatisfaction with disaster-related matters more frequently aired their views in answering the open question than those with a more positive or neutral view. This may have negatively biased the results.

APPENDIX 3

Sources consulted

The documents available are ordered alphabetically by department and institution. The annotation attempts to follow the original Dutch nomenclature and dating as far as possible.

Academic Medical Centre
- Research project, The prevention of post-traumatic stress disorder to victims and rescue workers as a result of the Bijlmer air disaster: a longitudinal study, 15/12/92.

Amsterdam City Police
- Log book Amsterdam City Police, 04/10/92 - 14/10/92.
- Evaluation regarding lists and identification.
- Log book police district 7 (Flierbosdreef), 4/10/92 - 14/10/92.
- Log book police Bijlmer Sports Centre, 4/10/92 - 10/11/92.
- Evaluation police district 7 (Flierbosdreef), 11/10/92.
- Lecture on crisis management Bijlmer Disaster, B.J.A.M. Welten.

Amsterdam City Printers
- Evaluation counter-disaster measures, 20/11/92.

Amsterdam Fire Brigade
- Evaluation counter-disaster measures fire service, 4/10/92 - 28/10/92.
- Regional fire service emergency centre, IJ-tunnel fire station, 12/10/92.

City Clerk's Office
- Evaluation counter-disaster measures, 27/11/92.

City General Administration and Legal Affairs Department (GALA)
- Evaluation counter-disaster measures of the GALA Department, 16/11/92.
- Report on the evaluation activities GALA Department during the disaster period, 14/10/92.

City Housing Department
- Report, Air disaster in the Bijlmermeer, by City Housing Department Director Denijs.

City Insurance Department
- Evaluation counter-disaster measures, 27/11/92.

City Medical Services Department

Abridged evaluation, Air disaster Zuidoost 04-10-1992, 3/11/92.

Comprehensive log book and remarks with reference to the evaluation air disaster Zuidoost 04-10-1992, 15/11/92.

Report after-care team city Medical Services Department, Air disaster Bijlmer, 30/11/93.

City Population Register

Evaluation counter-disaster measures, 13/10/92.

City Social Services Department

Log book Co-ordination Centre city Social Services Department, 6/10/92 – 11/10/92.

Log book crisis centre city Social Services Department, 5/10/92 – 21/10/92.

Evaluation report on (disaster) relief air disaster in the Bijlmer from 04 October 1992 up to and including 21 December 1992, February 1993.

City Public Transport Company

Evaluation and activity schemes, 17/12/92.

Co-ordination Centre

Log book Co-ordination Centre, 4/10/92 – 23/10/92.

Appendix log book Co-ordination Centre.

City Council disaster contingency plan Amsterdam, 30/11/88.

National Disaster Identification Team

The role of the Disaster Identification Team in large-scale traffic accidents, L.C. van der Pols.

Dutch Red Cross, Amsterdam Branch

Deployment report, Bijlmer air disaster.

Syllabus one-day workshop Provincial Command North Holland, 20/2/93.

Energy Company Amsterdam

Log book counter-disaster measures, 12/11/92.

Evaluation counter-disaster measures, 1/12/92.

Media

Various national and local news media.

Media surveys (parts 1 to 4) of the Information Board of the Ministry of Transport and Public Works.

Press, Information and Public Relations Department newspaper cuttings.

Press reports of the Press, Information and Public Relations Department and various city departments.

Audiovisual documentation.

Nieuw Amsterdam Housing Co-operative

Evaluation air disaster.

1991 annual report.

Press, Information and Public Relations Department

Evaluation air disaster Bijlmer, December 1992.

Information press conferences, 6/10/92 – 20/10/92.

Province of North Holland
Survey of facts and circumstances from the Provincial Crisis Centre.
Southeast Neighbourhood Council
District Zuidoost: relief workers on the front line, February 1993.
Steering Committee After-Care Bijlmer Air Disaster
Report from the Steering Committee After-Care Bijlmer Air Disaster, February 1993.
Work Group Air Traffic Bijlmer
Data file January 1993.

APPENDIX 4

Interviewees

Within the framework of this research around fifty interviews were held with the authorities and officials of the organisations, departments and institutions listed below:

Amsterdam City Hall
- Mayor
- General Executive and Legal Affairs Department
- Press, Information and Public Relations Department
- City Clerk's office
- various Aldermen

Bijlmer Sports Centre
City Housing Department
City Medical Services Department
City Population Register
City Social Services Department
Disaster Identification Team
Dutch Red Cross, Amsterdam Division
Fire Brigade
- Fire Brigade Amsterdam
- Fire Brigade Hilversum

Interest Groups
- Kwakoe Youth Centre
- FORSA
- 'Ghana Co-ordinating Committee'
- Churches' Liaison Council

Nieuw Amsterdam Housing Co-operative
Ministry of Internal Affairs
Ministry of Justice

Police
- Marnixstraat Headquarters
- Ganzenhoef Station
- District Station Flierbosdreef

Province of North Holland
Psychiatric Division of the Amsterdam Medical Centre
Public Prosecutor
Royal Netherlands Navy, Amsterdam Naval Base
Southeast Neighbourhood Council

Crisis Research Center

This is a publication of the Crisis Research Center (CRC) of the Joint Program of Public Administration of the University of Leiden and Erasmus University, Rotterdam, the Netherlands. CRC focuses its research activities on extraordinary circumstances, e.g. social conditions and developments which deviate sharply from everyday conceptions of order, and, therefore, generally solicit non-routine reactions in the community and on the part of government officials and agencies.

Whenever possible, CRC engages in quick-response research: making its own on-the-spot observations at the moment when information suggests that extraordinary circumstances have occurred. As soon as possible, CRC gathers additional data on the events and on the crisis management process. Within this overall research strategy, attention is focused on administrative and operational decision making within government, as well as activities of other stakeholders and the interaction between governmental and other organisations.

CRC's activities serve both academic and practical aims. The academic objective involves learning more about the similarities and the differences between extraordinary circumstances. This has resulted in many research publications on various aspects of crisis management, as well as in the founding and co-editing of the *Journal of Contingencies and Crisis Management* published by Basil Blackwell Ltd, which serves as an international and inter-disciplinary platform for findings, insights, and trend reports in this area. The practical objective manifests itself in an approach directed towards applications of CRC research findings. Key elements of that approach are: writing reports which are also accessible to those directly involved; making realistic recommendations; reporting as soon as possible; developing and conducting custom-built simulations; developing and conducting courses on crisis management; participating in public debates about social and technological vulnerabilities, social conflict, contingency planning, and crisis response. In addition, CRC regularly consults with governmental and other organisations and institutions both in and outside of the Netherlands on various aspects of crisis management.

Recent CRC publications

Rosenthal, U., Charles M.T. and Hart P. 't (Eds), (1989). *Coping with Crises: The Management of Disasters, Riots, and Terrorism*, Charles Thomas, Springfield, IL.

Rosenthal, U., Pijnenburg, B. (Eds), (1991). *Crisis Management and Decision Making: Simulation Oriented Scenarios*, Kluwer Academic Publishers, Dordrecht.

Journal of Contingencies and Crisis Management, Oxford: Basil Blackwell (quarterly, journal, co-edited by U. Rosenthal, together with Prof. A. Kouzmin, University of Western Sydney).

For Product Safety Concerns and Information please contact our EU representative GPSR@taylorandfrancis.com Taylor & Francis Verlag GmbH, Kaufingerstraße 24, 80331 München, Germany

Batch number: 10725802

Printed by Printforce, the Netherlands